Nanomaterials in Biological Systems
Interactions Between Nanoparticles and Macromolecules

Edited by

Jorddy N. Cruz

Center for Biological and Health Sciences, Department of Morphophysiology and Physiological Sciences, State University of Pará, Belém, Pará, Brazil

Copyright © 2026 by the authors

Published by **Materials Research Forum LLC**
Millersville, PA 17551, USA

Published as part of the book series
Materials Research Foundations
Volume 185 (2026)
ISSN 2471-8890 (Print)
ISSN 2471-8904 (Online)

Print ISBN 978-1-64490-384-1
eBook ISBN 978-1-64490-385-8

Distributed worldwide by

Materials Research Forum LLC
105 Springdale Lane
Millersville, PA 17551
USA
https://mrforum.com

Manufactured in the United States of America
10 9 8 7 6 5 4 3 2 1

Table of Contents

Preface

Nanotechnology stands at the vanguard of scientific innovation, redefining the boundaries of what is possible within medicine, biology, and environmental science. By manipulating matter at the nanoscale, researchers have unlocked unprecedented potential for precision, efficiency, and adaptability—ushering in an era where therapies can be tailored, diseases intercepted, and biological systems understood with extraordinary depth. Yet, with every stride toward progress arises a parallel obligation to examine the implications of these advances for health, ethics, and the environment.

This volume, *Nanoparticles in Biological and Environmental Systems*, brings together a collection of works that collectively illuminate the multifaceted nature of nanoparticle research and its profound implications for life sciences.

Chapter 1, *Nanoparticle-Based Approaches for Targeted Drug Delivery*, introduces the transformative capacity of nanocarriers designed for precise therapeutic targeting. It delineates how nanoparticles can overcome physiological barriers—such as the blood–brain barrier—thereby revolutionizing the treatment of complex diseases and cancers through enhanced delivery efficiency and biocompatibility.

Chapter 2, *Studies of Nanoparticles Applications in the Woman Body*, extends this perspective by exploring the physiological interactions and biomedical uses of nanoparticles specific to female biology. Through case studies, it underscores both their promise and the critical need to understand toxicological effects and long-term safety.

Revisiting the core theme, **Chapter 3** deepens the discussion on *Targeted Drug Delivery*, providing further mechanistic insights into the engineering of nanoparticles with tunable size, shape, and surface chemistry for optimized clinical outcomes.

Chapter 4, *Nanoparticle-Based Approaches for Wound Healing*, highlights the evolution of wound-care materials—from traditional antiseptics to nanostructured hydrogels—showcasing how nanoscale innovations enhance stability, bioavailability, and therapeutic response in tissue repair.

The dialogue progresses in **Chapter 5**, *Nanoparticles Interaction with Cancer Cells*, where the intricate interplay between engineered nanomaterials and the tumor microenvironment is examined. This chapter elucidates how magnetic, polymeric, and carbon-based nanoparticles contribute to precise diagnosis, targeted therapy, and the reimagining of cancer treatment paradigms.

In **Chapter 6**, *Organic Nanoparticles for Anti-Inflammatory Studies*, attention turns to the molecular mechanisms underpinning inflammation. Here, organic nanoparticles emerge as biocompatible and biodegradable tools capable of delivering anti-inflammatory agents efficiently, heralding new therapeutic opportunities for chronic inflammatory diseases.

Broader ecological dimensions are addressed in **Chapter 7**, *Environmental Impacts of Nanoparticles*. While nanomaterials fuel innovation across industries, their persistence, reactivity, and uncertain fate in ecosystems invite rigorous assessment. This chapter confronts the duality of nanotechnology's promise for environmental remediation and its potential ecological risks.

Finally, **Chapter 8**, *Ethical Considerations for Nanotechnology Applications in Biology*, closes the volume with a reflection on governance, equity, and responsibility. As nanotechnology permeates health and environmental domains, ethical discourse must guide its trajectory—ensuring transparency, justice, and respect for human and ecological integrity.

Together, these chapters weave a narrative that is both visionary and cautionary. They celebrate the ingenuity driving nanoscale science while advocating for a balanced approach—where innovation proceeds hand in hand with ethical mindfulness and environmental stewardship. This compendium aspires to serve not merely as a repository of knowledge, but as a call to harmonize technological advancement with the delicate complexity of life itself.

Nanomaterials in Biological Systems
Materials Research Foundations 185 (2026) 1-34

Materials Research Forum LLC
https://doi.org/10.21741/9781644903858-1

Chapter 1

Nanoparticle-Based Approaches for Reduced Immunity

Jibanananda Mishra[1]*, Jiban Jyoti Panda[2]

[1]School of Biosciences, RIMT University, Mandi Gobindgarh, Punjab 147301, India

[2]Chemical Biology Unit, Institute of Nano Science and Technology, Mohali -140306, Punjab, India

*mjiban@gmail.com

Abstract

Our immune system performs a significant role in defending our body against countless pathogens, but dysregulated or hyperactive immune responses can lead to chronic inflammatory disorders, autoimmune diseases, transplant rejection, and increased susceptibility to numerous infections. Nanoparticle (NP)-based strategies for reduced immunity represent a fast-evolving field with the potential to transform the treatment of immune-related disorders and improve patient outcomes. To address these challenges, various researchers have exploited the distinctive properties of nanoparticles (NPs) to fine-tune and modulate the immune system's activities. Meanwhile, various NP-based approaches, including nanomaterials, nanocarriers, and nanotherapeutics, are designed to achieve reduced immunity. These approaches involve a variety of strategies, ranging from immunosuppression for organ transplantation and autoimmune disease management to the prevention of unwanted inflammation in response to infection or injury. In this article, we have explored the intricate mechanisms and promising stratagems that have transpired at the juncture of nanotechnology and immunology.

Keywords

Nanoparticle, Immune System, Nanotherapeutics, Mechanism, Immunotherapy

Contents

1.   Introduction

The immune system stands as one of the most complex and finely-tuned biological networks of cells, proteins, and tissues in the human body [1]. In the continually evolving terrain of medical science, the intricate interplay between the human immune system and various diseases has been a topic of intense investigation. The human immune system plays a critical role in maintaining homeostasis and possesses an amazing defense mechanism, honed by evolution to protect the body from a multitude of external threats, including pathogens, toxins, and malignant cells [2]. However, this same immune system can sometimes turn against the body, leading to the development of autoimmune diseases, chronic inflammation, increased susceptibility to infections, or exacerbating the damage caused by certain diseases. While a robust immune response is essential for maintaining health, there are instances where immune reactions need to be curtailed, either to mitigate the harmful effects of autoimmune disorders, prevent organ transplant rejection, or control inflammation associated with chronic diseases [3]. Achieving this delicate balance between a fully active and subdued immune system has been a longstanding challenge in the realm of medicine.

NPs (NPs) have exhibited remarkable potential in a myriad of applications in various biomedical arenas, such as imaging, biosensing, gene and drug delivery. However, the influence of NPs on the immune system with respect to immunomodulation, is an area of research that has gained substantial prominence [4,5]. The NP-based approaches for modulating the immune system in various biomedical disciplines have surfaced as promising tools due to their distinctive physicochemical properties, such as tunable size and shape, which allow for precise control over their interactions with biological systems [6]. These nanomaterials have the capacity to interact with immune cells, such as macrophages, dendritic cells, and lymphocytes, in ways that can either augment or diminish their activities. The ability to manipulate these interactions unveils a vast array of opportunities for therapeutic interventions, ushering in an era where NPs serve as immunomodulatory agents [7].

This chapter aims to provide a comprehensive overview of the evolving landscape of NP-based approaches for reduced immunity. The exploration of these innovative strategies promises to unlock new avenues for precise and targeted immune modulation, potentially revolutionizing the treatment of autoimmune diseases, transplantation, cancer, and beyond. While the road ahead may be paved with challenges and uncertainties, the potential benefits for patients and the field of medicine are profound and exciting. As we explore deeper into this promising field, we hope to shed light on the remarkable strides being made at the crossroads of nanotechnology and immunology.

2. Types of NPs used for immune modulation

Different types of NPs have been investigated for their immunomodulatory properties and their ability to interact with immune cells. Here are some types of NPs commonly used for immune modulation:

2.1 Metallic NPs

Metallic NPs, such as gold, silver, iron oxide, titanium dioxide NPs have been generally studied for their immunomodulatory effects. These NPs have shown potential applications in cancer immunotherapy and other immunomodulatory treatments [8–10]. Metallic NPs can interact with immune cells such as T cells, macrophages, dendritic cells, and modulate immune responses [11]. Gold NPs, in particular, have been used as adjuvants in vaccine development to enhance immune responses [12]. Titanium dioxide (TiO_2) and silver NPs have been studied for their immunomodulatory effects [9]. Exposure to TiO_2 NPs has been shown to increase reactive oxygen species (ROS) levels in peripheral blood mononuclear cells (PBMCs) [9]. Additionally, exposure to these NPs can elicit systemic T lymphocyte activation and stimulate proinflammatory cytokine production by T lymphocytes. Silver NPs have similar effects on neutrophils and macrophages as ionic silver [9]. Furthermore, Yang et al. (2019) have shown that titanium NPs have been found to activate proinflammatory macrophages and increase proinflammatory cytokine secretion; however, the addition of lithium chloride has been found to suppress the inflammatory response stimulated by titanium NPs [13]. Another study by Amina et al. (2020) has shown that biogenic gold and silver NPs could reduce the pro-inflammatory cytokine levels in human leukemic monocyte cells (THP1) and natural killer cells (NK92) [14]. NPs can also be used in combination with other therapies to enhance their immunomodulatory effects. For example, Li et al. (2020) have shown that near-infrared (NIR) photoactivatable immunomodulatory black phosphorus nanosheets and quantum dots grafted with polyethylene glycol (PEG) and ROS-sensitive polypropylene sulfide holds a great promise in cancer immunotherapy. These NPs can be activated by NIR light to induce immunomodulatory effects in the tumor microenvironment with deeper penetration [15]. NPs can be used as adjuvants in immunotherapy due to their biocompatibility and tailored physicochemical properties. They can enhance the delivery, retention, and release of immunostimulatory agents and biologicals in targeted cell populations and tissues [4]. Moreover, NPs can be functionalized to mimic self-markers, which can improve their biocompatibility and reduce phagocytosis [16].

2.2   Lipid-based NPs

Lipid-based NPs have appeared as promising tools for immunomodulation and in various biomedical applications, including cancer immunotherapy [10,17,18]. These NPs can encapsulate antigens or immunomodulatory molecules such as microRNAs (miRNAs) and efficiently target and deliver them to immune cells, triggering specific immune responses [11] [10]. For instance, Lin *et al.*, (2020) have designed lipid-coated calcium phosphonate NPs to deliver microRNAs for immunomodulation in cancer immunotherapy. Lipid-based NPs have shown potential in vaccine development and immunotherapy too [11].

Lipid-based NPs offer advantages in the selective and localized delivery of immunomodulators, tumor-associated antigens (TAAs), and drugs to targeted tumor sites or lymphoid tissues [18]. They can enhance the amount of antigen reaching immune cells and improve the efficacy of immunotherapy [19]. Additionally, lipid-based nanocarriers can be used for the delivery of Toll-like receptor agonists, which can activate the immune system and enhance the immune response [20].

In personalized cancer vaccines (PCVs), lipid-based NPs can be utilized to modulate the innate and adaptive immune systems by delivering customized tumor-specific antigens [17]. These NPs can activate antitumor immunity and improve the efficacy of cancer immunotherapy [17]. Furthermore, lipid-based nanocarriers have been explored for the delivery of immunomodulators, such as TLR7/8 agonists and SHP2 inhibitors, to enhance macrophage immunotherapy [21]. The immunomodulatory effects of lipid-based NPs are not limited to cancer immunotherapy. They have also been investigated in the treatment of inflammatory bowel disease (IBD). Lipid nanocarriers loaded with curcumin have shown anti-inflammatory properties and the ability to reduce colitis symptoms [22]. It is worth noting that the choice of lipid composition and NP design can influence the immunomodulatory properties of lipid-based NPs. For example, the surface charge and size of NPs can affect their interactions with immune cells and the immune response [19]. Additionally, the incorporation of specific immunomodulatory agents or adjuvants into lipid-based NPs can further enhance their immunomodulatory effects [17].

In short, lipid-based NPs hold great potential for immunomodulation in various applications, including cancer immunotherapy and the treatment of inflammatory diseases. These NPs can efficiently deliver immunomodulatory agents to immune cells and enhance the immune response. However, further research is needed to optimize their design and understand their mechanisms of action for effective immunomodulation.

2.3   Polymer-based NPs

Polymer-based NPs have shown great potential for immunomodulation in various applications, including vaccine delivery systems and cancer immunotherapy [10,23,24]. These NPs can be engineered to modulate their physical-chemical properties, such as size, composition, surface charge, and flexibility, to enhance their immunomodulatory effects [25].

Polymer-based NPs, including polymeric micelles, have been investigated for their immunomodulatory effects. These NPs can be engineered to carry antigens, immunomodulatory agents, or drugs, and deliver them to immune cells. Polymer-based NPs have shown promise in cancer immunotherapy and vaccine development [11].

In the field of vaccine delivery systems, polymer-based NPs have been extensively studied for their ability to enhance the immunogenicity of subunit vaccines [23]. These NPs can improve the delivery of antigens and immunostimulatory agents, leading to enhanced immune responses. For instance, poly(γ-glutamic acid)-graft-L-phenylalanine ethyl ester (γ-PGA-Phe) NPs have been used to encapsulate CpG, a danger signal that activates the immune system, resulting in improved immunogenicity [23]. Additionally, the surface chemistry of NPs, such as disulfide-sensitive moieties, can influence the activation of the complement cascade and the initiation of adaptive immune responses [23].

Polymer-based NPs have also been explored in cancer immunotherapy. They can efficiently deliver microRNA-based therapeutics to immune cells for immunomodulation [10]. These NPs can target immune cells and modulate their functions, leading to enhanced antitumor immune responses. Furthermore, RNA molecules, such as siRNA, miRNA, and mRNA, delivered by polymer-based NPs, have shown potential for immunomodulation and cancer immunotherapy [10]. One important aspect of polymer-based NPs is their ability to induce innate immune memory in monocytes and macrophages [24,26]. These NPs can prime macrophages for a more potent response to subsequent stimuli and modulate bacteria-induced memory. This capacity of engineered NPs opens up new possibilities for immunomodulation, including improving immune responses to vaccines, enhancing resistance to infections, and modulating immune and inflammatory reactions in chronic inflammatory diseases and autoimmunity [24].

Briefly, polymer-based NPs offer a versatile platform for immunomodulation in various applications. They can enhance the immunogenicity of vaccines, deliver immunomodulatory agents for cancer immunotherapy, and induce innate immune memory in monocytes and macrophages. However, further research is needed to optimize the design and formulation of polymer-based NPs for effective immunomodulation.

2.4   Carbon-based NPs

Carbon-based NPs, such as carbon nanotubes and graphene oxide NPs, have also been studied for their immunomodulatory properties. These NPs can interact with immune cells and modulate immune responses. Carbon-based NPs have shown potential in drug delivery and immunotherapy applications too [11]. Inoue *et al.* (2006) demonstrated that carbon NPs can facilitate antigen-related airway inflammation in mice [27]. Similarly, Huizar *et al.* (2011) observed acute inflammation and pulmonary granuloma-like lesions in mice exposed to aggregated carbon NPs [28]. These findings suggest that carbon NPs can induce immune responses and inflammation in the lungs. Furthermore, carbon NPs have been shown to modulate immune cells. Huq *et al.* (2016) found that carbon NPs (polyethylene glycol-functionalized hydrophilic carbon clusters) preferentially target T lymphocytes, indicating their potential as immunomodulators [29]. In addition, Zhang *et al.* (2021) found that carbon NPs can activate dendritic cells (DCs) through a pathogen-mimicking behavior. DCs play a crucial role in initiating immune responses, suggesting that carbon NPs can influence immune cell function [30].

The immunomodulatory properties of carbon NPs have also been explored in cancer immunotherapy. Nguyen *et al.* (2018) have discussed the use of mesoporous silica NPs as a platform for cancer immunotherapy [31]. These NPs can deliver immune-modulating molecules to target sites such as lymph nodes and tumor sites, enhancing the therapeutic efficacy of immunotherapy. Similarly, Yang *et al.* (2020) highlighted the importance of optimizing the structural properties of MnO_2 NPs for their immunomodulatory activity in cancer starvation-immunotherapy [32].

Carbon NPs have also been studied in the context of allergic reactions. Kroker *et al.* (2015) investigated the role of dendritic cells in carbon NP-induced allergic airway inflammation. They found that animals exposed to antigen in the presence of carbon NPs showed increased effects with respect to ovalbumin sensitization and allergic airway inflammation [33]. This suggests that carbon NPs can enhance allergic sensitization and subsequent allergic reactions.

Carbon nanotubes (CNTs) have also shown promise in immunomodulation. CNTs possess unique physical properties, including electrical, thermal, and spectroscopic properties, which make them advantageous for disease detection and therapy [34]. The interaction between carbon-based NPs and amino acids has also been investigated, revealing the potential for molecular dynamics simulations to study these interactions [35–37].

In summary, carbon NPs have been shown to have immunomodulatory effects in various studies. They can induce immune responses and inflammation in the lungs, modulate immune cell function, and have potential applications in cancer immunotherapy and allergic reactions. Further research is needed to fully understand the mechanisms underlying the immunomodulatory properties of carbon NPs and their potential therapeutic applications.

2.5   Inorganic NPs

Inorganic NPs, such as silica NPs and quantum dots, have been explored for their immunomodulatory effects. These NPs can be functionalized to interact with immune cells and modulate immune responses. Inorganic NPs have shown potential in vaccine development and immunotherapy [11].

Inorganic NPs have gained significant attention in the field of immunomodulation due to their unique properties and potential applications in cancer therapy and vaccine development [38,39]. These NPs can serve as delivery platforms for immunomodulators, enhancing anti-tumor immune responses and providing more precise and effective tumor treatment [38]. Various types of inorganic NPs, such as mesoporous silica NPs (MSNs), gold-based nanomaterials, TiO_2 NPs, silver NPs, and selenium NPs, have been investigated for their immunomodulatory effects [39–41].

One of the key advantages of inorganic NPs is their shape-controllable and size-adjustable attributes, which allow for precise control over their physicochemical properties and drug delivery capabilities [39]. For example, gold NPs have been used as adjuvants to increase the effectiveness of vaccines by stimulating antigen-presenting cells and ensuring controlled antigen release [42]. Similarly, TiO_2 NPs and silver NPs have been shown to induce immunomodulatory effects, including activation of T lymphocytes and proinflammatory cytokine production [9]. Selenium NPs have also attracted attention due to their higher bioactivity and lower toxicity compared to other selenium species [40]. In addition to their role as delivery platforms, inorganic NPs themselves can directly modulate the immune response. For instance, silica/polyphosphate NPs have been used to inhibit the binding of viral proteins to host cells, thereby preventing viral infection [43]. Furthermore, NPs can be designed to target specific immune cells or organs, such as the spleen, which plays a crucial role in regulating the systemic immune system [44]. NPs targeted to the spleen can enhance immunomodulation and provide systemic immune regulation. Overall, inorganic NPs offer great potential for immunomodulation in various applications, including cancer therapy, vaccine development, and the treatment of inflammatory diseases. Their unique properties, such as shape controllability, size adjustability, and targeting capabilities, make them promising candidates for enhancing immune responses and improving therapeutic outcomes. Further research is needed to fully understand the mechanisms of action and optimize the design of inorganic NPs for immunomodulation.

Carbon-based NPs have shown potential for immunomodulation in various biomedical applications, including inflammation therapy and cancer immunotherapy [10,45,46]. These NPs, such as CNTs and graphene oxide (GO), possess unique chemical and physical properties that make them suitable for biomedical applications.

In the context of inflammation therapy, cell membrane-based nanotherapeutics coated with cell membranes derived from cells involved in the inflammatory process have been investigated [45]. These nanotherapeutics can target and manipulate the inflammatory microenvironment, modulating the immune response [45]. Additionally, carbon black NPs have been implicated in the induction of inflammasome-dependent pyroptosis, a form of programmed cell death, in macrophages. Inhalation of carbon black NPs has been associated with respiratory morbidity and mortality [47].

In the field of cancer immunotherapy, carbon-based NPs have been explored for their potential in delivering RNA-based therapeutics. RNA molecules, such as siRNA, miRNA, and mRNA, have shown promise for immunomodulation and cancer immunotherapy by activating innate and adaptive immune responses [10]. NP platforms, including carbon-based NPs, have been utilized to enhance the delivery of RNA to tumor and immune cells.

Furthermore, metallic NPs, including carbon-based NPs, have been investigated for their antiviral potential against viral infections, including COVID-19 [48]. These NPs can act as carriers for the targeted delivery of antiviral molecules and as diagnostic agents for the rapid and sensitive diagnosis of viral infections. It is important to note that while carbon-based NPs hold promise for immunomodulation, there are also concerns regarding their potential adverse effects and environmental contamination [47]. However, further research is needed to fully understand the interactions between carbon-based NPs and the immune system, as well as to optimize their design and translation for clinical applications.

2.6   Hybrid NPs

Hybrid NPs, which combine different types of materials, have also been investigated for immune modulation. These NPs can offer synergistic effects and enhanced immunomodulatory properties [11]. Hybrid NPs have shown potential in various biomedical applications, including drug delivery and immunotherapy [11]. Besides, hybrid NPs have demonstrated significant potential in immunomodulation and cancer therapy. These NPs are designed by combining different materials and functionalities to target specific cells or tissues, deliver therapeutic agents, and modulate the immune response [49]. They offer several advantages, including improved drug delivery efficiency, regulation of the immunosuppressive tumor microenvironment, activation of antitumor immunity, and enhanced antitumor effects [50]. One example of hybrid NPs used in immunomodulation is the combination of TiO_2 and silver NPs. TiO_2 NPs have been shown to increase ROS levels in PBMCs post-exposure [9]. This surge in ROS levels can have immunomodulatory effects on immune cells. Additionally, silver NPs have been found to possess immunomodulatory properties and can inhibit metastasis formation and prolong survival rates in a liver metastatic model of breast cancer [51]. Another approach to immunomodulation using hybrid NPs is the incorporation of immunostimulatory molecules into RNA NPs. For instance, RNA NPs incorporating CpG oligodeoxynucleotide have been shown to activate innate immune responses [52]. These NPs can deliver CpG and short interfering RNA (siRNA) to immune cells, leading to the activation of immune responses and the modulation of the tumor microenvironment.

Lipid-polymer hybrid NPs have also been utilized for immunomodulation in cancer therapy. These NPs can deliver therapeutic agents and immunostimulatory molecules to immune cells, resulting in the activation of antitumor immunity and tumor eradication [53]. They can specifically target the tumor microenvironment and effectively downregulate vascular endothelial growth factor (VEGF), which is involved in tumor angiogenesis [53].

In addition to delivering therapeutic agents and immunostimulatory molecules, hybrid NPs have been found to enhance the activity of immune cells. Magnetic NPs (MNPs), for example, have been reported to stimulate the secretion of extracellular vesicles (EVs) by immune cells, which can be utilized for the rapid production of EVs for immunotherapy [54]. MNPs can also enhance the activity of natural killer (NK) cells, which play a crucial role in cancer immunotherapy [55,56]. NPs can enhance NK cell activity, promote the homing of NK cells to tumors, deliver RNA interference (RNAi) to enhance NK cell activity, and genetically modulate NK cells based on NPs [55].

Furthermore, hybrid NPs can be designed to target specific cells or tissues for immunomodulation. Cell membrane-coated NPs, for instance, have been developed for cancer immunotherapy. These NPs combine the membrane of immune cells, such as dendritic cells (DCs), with the membrane of cancer cells. The cancer cell membrane provides tumor antigens, while the DC membrane provides costimulatory signals, leading to a stronger anti-cancer immune response [57]. Similarly, mesenchymal stem cell (MSC)-derived exosomes packaged with magnetic NPs have been used for targeted tumor cell ablation via magnetically induced hyperthermia. These exosomes can be efficiently endocytosed by tumor cells, resulting in targeted tumor cell ablation [58].

Thus, hybrid NPs offer a promising approach to immunomodulation and cancer therapy. They can deliver therapeutic agents, activate immune responses, modulate the tumor microenvironment, enhance the activity of immune cells, and target specific cells or tissues. Further research in this field is necessary to optimize the design and application of hybrid NPs for immunomodulation and cancer therapy.

2.7   Quantum dots (Qdots)

Qdots are fluorescent semiconductor nanocrystals that have gained significant attention in the field of immune modulation. They offer unique optical properties, such as high brightness and photostability, making them valuable tools for various applications in immunology [59]. One area of research involving quantum dots is their use in live cell imaging and diagnostics. Qdots have been utilized for high-resolution cellular imaging, allowing for the visualization of intracellular processes at the single-molecule level. They can be functionalized with specific ligands or antibodies to target specific immune cells or biomarkers, enabling the detection and characterization of immune responses [59].

Furthermore, Qdots have been explored as a platform for NP drug delivery (NDD) systems. NDD has emerged as a promising approach to improving the efficacy of existing drugs and enabling the development of new therapies [60]. Qdots can be incorporated into NDD vehicles, maintaining their properties and behavior while providing mechanisms for monitoring intracellular and systemic nanocarrier distribution. The small size and versatile surface chemistry of Qdots allow for their incorporation within various NDD vehicles with minimal effect on overall characteristics [60]. However, it is important to consider the safety aspects of Qdots in immune modulation. There is growing concern about the safety of engineered NPs, including Qdots, due to their potential to induce oxidative stress and cell death. Studies have shown that Qdots can induce changes in intracellular metabolism and the accumulation of lipids without compromising cellular viability [61]. Therefore, further research is needed to fully understand the potential toxicological effects of Qdots and to develop strategies for their safe use in immune modulation.

Briefly, Qdots have shown great potential in immune modulation, particularly in live cell imaging, diagnostics, and drug delivery systems. Their unique optical properties and versatile surface chemistry make them valuable tools for studying immune responses and developing targeted therapies. However, the safety aspects of Qdots should be carefully considered and further investigated to ensure their safe and effective use in immunological applications.

2.8   Peptide NPs

Peptide NPs have shown potential in immune modulation and immunotherapy. They can be engineered to have specific physical-chemical properties that allow for targeted delivery of immunomodulatory agents, such as cytokines or immune checkpoint inhibitors [24,62,63]. One important aspect of peptide NPs in immune modulation is their ability to activate cells of the innate immune system, such as monocytes and macrophages. These NPs can induce inflammatory activation and promote immune responses. They can also induce innate immune memory in monocytes and macrophages, leading to enhanced immune responses upon subsequent encounters with pathogens or antigens [24]. Furthermore, peptide NPs can be used to deliver cytokines and other immunomodulatory agents to specific immune cells, such as CD4+ T lymphocytes [62]. This targeted delivery allows for precise modulation of immune responses and can be utilized in the treatment of various diseases, including cancer and autoimmune disorders.

In the context of cancer immunotherapy, peptide NPs can be used to deliver immune checkpoint inhibitors, such as anti-PD-1 or anti-CTLA-4 antibodies, to tumor sites [63]. This targeted delivery enhances the efficacy of immunotherapy by blocking inhibitory signals and promoting anti-tumor immune responses. Peptide NPs can also be used to deliver cytokines, such as interleukin-2 (IL-2) or interleukin-12 (IL-12), to enhance the activation and proliferation of immune cells [63]. Moreover, peptide NPs can be functionalized with specific ligands or antibodies to target immune cells or tumor cells, allowing for selective delivery of immunomodulatory agents [63]. This targeted approach minimizes off-target effects and improves the therapeutic index of immunotherapies.

The use of peptide NPs in immune modulation and immunotherapy is still an area of active research. Further studies are desired to optimize the design and formulation of peptide NPs, understand their interactions with the immune system, and evaluate their safety and efficacy in preclinical and clinical settings.

In short, peptide NPs have shown promise in immune modulation and immunotherapy. They can be engineered to deliver immunomodulatory agents to specific immune cells, enhance immune responses, and improve the efficacy of immunotherapies. The targeted delivery and modulation of immune responses by peptide NPs hold great potential for the treatment of various diseases, including cancer and autoimmune disorders.

3.   Mechanisms of immune suppression by NPs

The mechanisms of immune modulation by NPs have been extensively studied by researchers. One mechanism involves the anti-inflammatory properties of NPs, such as CNTs, which have been shown to suppress B-cell function [64]. Another mechanism involves the production of transforming growth factor-beta (TGF-β) by alveolar macrophages, which plays a role in immunosuppression. Metal-based NPs have also been found to interact with the immune system and trigger immune responses. These NPs can activate both the innate and adaptive immune responses. They can induce inflammation and cause immunotoxicity, leading to pathological changes in the immune system. However, some metal-based NPs have been shown to be immunocompatible and do not interfere with the immune response [11].

NPs can also interact with specific immune cells, such as dendritic cells (DCs). Studies have shown that NPs can stimulate DC maturation and antigen presentation, which are crucial for the initiation of effective immune responses [42,65]. Gold NPs, in particular, have been used as adjuvants to increase the effectiveness of vaccines by stimulating antigen-presenting cells and ensuring controlled antigen release [42]. Furthermore, NPs can serve as delivery systems for antigens, either by being engulfed by immune cells or by releasing antigens at the target location [66]. This controlled and sustained release of antigens, along with other immune modulators, can enhance immune responses and promote lasting immune memory. NPs can also bind to bacterial membranes and preserve the biological properties of lipopolysaccharides for natural antigen presentation to immune cells, thereby inducing immune responses against infectious diseases [41].

It is important to consider the physicochemical properties of NPs, such as size, charge, hydrophobicity, and hydrophilicity, as they can influence cellular responses and immune reactions. The interaction of NPs with phagocytic cells and their mechanisms of intracellular penetration are also of interest in understanding the immune response to NPs [42].

Concisely, the mechanisms of immune modulation by NPs involve the anti-inflammatory properties of NPs, the production of immunosuppressive factors, the activation of immune responses by metal-based NPs, the stimulation of dendritic cell maturation and antigen presentation, and the use of NPs as antigen delivery systems. The physicochemical properties of NPs and their interactions with immune cells play a crucial role in determining the immune response. Further research is needed to fully understand these mechanisms and optimize the use of NPs in immunological applications.

3.1   Interactions of NPs with immune cells

The interaction of NPs with immune cells has been extensively studied. NPs can interact with various immune cells, including phagocytic cells, dendritic cells, macrophages, B cells, and T cells [41,42,64,67–70]. These interactions have significant effects on the immune response and can be utilized in the development of vaccines and immunotherapies. One important aspect of NP-immune cell interaction is the internalization of NPs by immune cells. Phagocytic cells, such as macrophages and dendritic cells, have the ability to engulf and eliminate NPs [67]. The mechanisms of NP internalization by immune cells, including the mechanisms of intracellular penetration, have been studied [42]. The physicochemical properties of NPs, such as size, charge, and shape, can influence their internalization by immune cells. For instance, NPs with a positive charge have been shown to have increased uptake by immune cells compared to negatively charged NPs [71].

The internalization of NPs by immune cells can lead to various immune responses. NPs can induce the production of cytokines, such as interferon-gamma (IFN-γ), and stimulate the activation of T cells [42,70]. They can also enhance antigen processing and presentation by antigen-presenting cells, such as dendritic cells, leading to the activation of B cells and the secretion of antibodies [42,72]. Furthermore, NPs can modulate the immune response by affecting the balance between pro-inflammatory and anti-inflammatory cytokines.

The biodistribution of NPs in the immune system is another important aspect to consider. NPs can interact with immune cells both in the bloodstream and in tissues. Manipulation of NP size and charge can influence their delivery to immune cells and their biodistribution [68]. For example, NPs can be engineered to specifically target immune cells or to avoid immune recognition [67].

The interaction of NPs with immune cells has implications for various applications, including drug delivery and vaccine development. NPs can be used as carriers for drugs and antigens, allowing for controlled and sustained release [72]. They can also be used as adjuvants to enhance the immune response to vaccines [42]. The physicochemical properties of NPs can be tailored to optimize their interaction with immune cells and improve their efficacy in these applications.

The interaction of NPs with immune cells is a complex process that involves internalization, immune response modulation, and biodistribution. The physicochemical properties of NPs play a crucial role in determining their interaction with immune cells. Understanding these interactions is essential for the development of NP-based therapies and vaccines.

3.2   Immunomodulatory properties of NPs

NPs have been found to possess immunomodulatory properties, allowing them to influence the immune response in various ways. They can both enhance and suppress immune responses, depending on the specific context and application [5]. In terms of immune suppression, NPs have shown potential in promoting immune tolerance in conditions such as chronic or acute inflammations, autoimmune diseases, transplant rejection, and allergies [5,73]. These diseases are characterized by inappropriate overreactions of the immune system, and NPs can help regulate and dampen these responses [73].

There are several mechanisms by which NPs induce immune suppression. One approach is through the delivery of antigens alone, which harnesses natural tolerogenic processes or environments [73]. NPs can also provide antigens while targeting pro-tolerogenic receptors, further promoting immune tolerance. Additionally, NPs can use specific strategies to initiate and modulate immune responses, such as targeting antigen-presenting cells and receptors involved in immunological processes [73]. Furthermore, NPs derived from edible sources have been found to regulate immune homeostasis in the intestine by targeting dendritic cells [74]. These NPs, such as those derived from broccoli, activate dendritic cell AMP-activated protein kinase (AMPK), which plays a role in preventing excessive immune activation and inflammation [74].

Nanomaterials in Biological Systems   Materials Research Forum LLC
Materials Research Foundations 185 (2026) 1-34   https://doi.org/10.21741/9781644903858-1

In the context of cancer immunotherapy, NPs have been utilized to overcome the immune-suppressive tumor microenvironment (TME) and enhance anti-cancer immunity. For example, NPs can be used to deliver cancer antigens and adjuvants, promoting immune responses against tumor cells [49]. They can also be used to locally activate T cells in the TME, sensitizing tumors to immune checkpoint blockade therapy [75].

NPs have demonstrated immunosuppressive effects by promoting immune tolerance and dampening excessive immune reactions. They achieve this through various mechanisms, including antigen delivery, targeting of pro-tolerogenic receptors, and modulation of immune cell activation. These immunomodulatory properties of NPs have potential applications in the treatment of autoimmune diseases, allergies, transplant rejection, and cancer immunotherapy.

3.4   Cytokine and chemokine regulation

The regulation of cytokines and chemokines by NPs has been investigated in various studies. These studies have explored the potential of NPs in modulating immune responses and treating diseases such as sepsis, cancer, and COVID-19. In a study by Selvaraj *et al.* (2015), cerium oxide NPs were shown to inhibit MAP kinase/NF-kB-mediated signaling and attenuate lipopolysaccharide-induced severe sepsis. The treatment with NPs resulted in the modulation of serum cytokine/chemokine levels, which are closely associated with patient survival [76]. Hou *et al.* (2018) investigated the use of NP-based photothermal and photodynamic immunotherapy for tumor treatment. They found that photothermal therapy using PLGA-ICG NPs, combined with the immunomodulator imiquimod, promoted cytokine secretion and inhibited tumor growth without significant immune-related adverse effects [77].

In the context of the COVID-19 cytokine storm, Meng *et al.* (2021) discussed the potential of advanced materials, including NPs, for capturing cytokines and managing inflammatory diseases. The review highlighted the use of emerging biomaterials for improving antibody-based and broad-spectrum cytokine neutralization [78]. Another study by Selvaraj *et al.* (2015) focused on the inhibition of MAP kinase/NF-kB-mediated signaling and attenuation of lipopolysaccharide-induced severe sepsis using cerium oxide NPs. The researchers observed that many of the cytokines/chemokines and other inflammatory regulators associated with systemic inflammatory response syndrome (SIRS) are likely derived from liver Kupffer cells [76]. Furthermore, in the study by Hou *et al.* (2018), NP-based photothermal and photodynamic immunotherapy for tumor treatment were investigated. The researchers found that the massive tumor cell death induced by photothermal therapy and photodynamic therapy triggered immune responses, including the expression and secretion of cytokines, activation of immune effector cells, and transformation of memory T lymphocytes [77].

These studies collectively demonstrate the potential of NPs in modulating cytokine and chemokine levels, promoting immune responses, and treating various diseases. The unique properties of NPs allow for targeted delivery of immunomodulatory agents and precise regulation of immune responses. However, further research is needed to optimize NP design, understand their interactions with the immune system, and evaluate their efficacy and safety in clinical settings.

3.5   T-cell and B-cell responses

T-cell and B-cell responses to NPs have been investigated in various studies. The interaction of NPs with immune cells can influence the activation and differentiation of T and B cells, leading to specific immune responses [23,79]. NPs can modulate T-cell responses by delivering antigens and immunostimulatory molecules to antigen-presenting cells (APCs). The composition, size, and surface charge of NPs can impact their biodistribution and the delivery of antigens to APCs, thereby influencing T-cell activation and immune responses. NPs can enhance the immunogenicity of vaccines by improving antigen presentation and promoting the activation of T cells [23].

The interaction between NPs and T-cell and B-cell responses has been investigated in several studies. In the case of B-cell responses, NPs have been shown to influence the generation and function of memory B cells [79]. Memory B cells play a vital role in serological immunity and can rapidly differentiate into plasma cells upon re-exposure to antigens. NPs can enhance the generation of memory B cells by modulating the activation of complement cascade and the initiation of adaptive immune responses. Besides, NPs can influence the affinity and isotype switching of B cells, leading to the production of antibodies with higher affinity and different immunoglobulin classes [79].

Furthermore, NPs have been shown to induce innate immune memory in monocytes and macrophages, which can impact subsequent immune responses [24]. The activation of innate immune cells by NPs can lead to the production of pro-inflammatory cytokines and the initiation of immune memory. This innate immune memory can enhance the responsiveness of immune cells upon subsequent encounters with pathogens or antigens [24].

Overall, NPs can modulate T-cell and B-cell responses by delivering antigens, immunostimulatory molecules, and influencing the activation and differentiation of immune cells. The composition, size, and surface properties of NPs play a crucial role in determining their immunomodulatory effects. Further research is needed to fully understand the mechanisms underlying the interaction between NPs and immune cells and to optimize NP design for immunotherapeutic applications.

The interaction between NPs and T-cell and B-cell responses has been investigated in several studies. In a study by Lenz *et al.* (2009), a dose-controlled system for air-liquid interface cell exposure was used to assess the effects of ZnO NPs on cellular responses. The study found that ZnO NPs did not induce a cellular response below a certain concentration, indicating that they were not toxic at occupationally allowed exposure levels [80]. This suggests that ZnO NPs may not have a significant impact on T-cell and B-cell responses. Another study by Zhu *NP* (2012) focused on TiO_2 NPs and their effects on cell death. The study found that TiO2 NPs preferentially induced cell death in transformed cells in a Bak/Bax-independent manner. This suggests that TiO_2 NPs may have the potential to modulate T-cell and B-cell responses by affecting cell viability and survival [81].

These studies provide insights into the potential effects of NPs on T-cell and B-cell responses. However, it is important to note that the specific effects may vary depending on the type, size, concentration, and surface properties of the NPs, as well as the specific immune cell types and experimental conditions used in each study. Further research is required to fully understand the mechanisms underlying the interaction between NPs and T-cell and B-cell responses and to evaluate the potential immunomodulatory effects of different types of NPs.

3.6 Targeting specific immune pathways

NPs can be designed to target specific immune cell signaling pathways, offering a potential strategy for modulating immune responses. The influence of NP properties, such as elasticity and shape, on immune cell interactions and signaling pathways has been investigated by Anselmo *NP* (2015). The authors conducted a study to investigate the impact of NP elasticity on their biological fate. The study demonstrated that tuning the elasticity of NPs can enhance their circulation time, reduce immune system uptake, and improve targeting capabilities [82]. This suggests that NP elasticity can influence immune cell interactions and signaling pathways, potentially leading to desired immunomodulatory effects. In the same study, the role of NP elasticity on various key functions, including blood circulation time, biodistribution, antibody-mediated targeting, endocytosis, and phagocytosis, was investigated [82]. The results showed that softer NPs with lower elasticity exhibited enhanced circulation and improved targeting compared to harder NPs [82]. This indicates that NP elasticity can impact immune cell interactions and signaling pathways, potentially affecting immune responses. Furthermore, Anselmo *NP* (2015) highlighted the shape of NPs as another physical parameter that can be tuned to improve their biological functions. The shape of NPs has been extensively studied and can influence their interactions with immune cells and signaling pathways [82]. Different NP shapes, such as spheres, rods, and discs, can elicit distinct immune responses and modulate immune cell functions.

Different NPs have been found to act on different signaling pathways in immune cells. For example, gold NPs have been shown to activate the NLRP3 inflammasome, which is involved in innate immune signaling. Activation of the inflammasome by NPs can enhance antibody production and promote an immune response [83]. The interaction between NPs and immune cells is complex and can have both beneficial and detrimental effects. On one hand, NPs can trigger inflammatory responses and potentiate innate immune responses, which can be beneficial for immunotherapy. On the other hand, NPs can induce toxicities, such as tissue fibrosis or allergy, through inflammatory responses [83]. The immunostimulatory effect of NPs is not fully understood, but it has been shown that NPs can induce the activation of the NLRP3 inflammasome and the release of cytokines [83,84]. Besides, heparosan-coated NPs have been found to be endocytosed by innate immune cells via clathrin-mediated and macropinocytosis pathways [71].

Targeting specific receptors on immune cells is another strategy to enhance NP uptake and immune response. Surface functionalization of NPs with specific ligands, such as chitosan and mannose, has been shown to deliver NPs to macrophages and dendritic cells via specific phagocytic pathways, improving the immune response to NP-bound antigens [68,85]. Similarly, camouflaging NPs with cellular membranes from immune cells, such as macrophages and dendritic cells, has emerged as a promising strategy for cancer theragnostics. The membrane-coated NPs retain membrane-bound antigens and other biologically relevant moieties, allowing for immune evasion and targeted delivery [86].

Nanotherapeutics, which are NPs designed for therapeutic purposes, have been shown to interact with immune cells and modulate signaling pathways involved in immune responses. These interactions can have significant effects on the biodistribution of NPs and their ability to deliver therapeutic payloads to immune cells [68]. One important aspect of nanotherapeutics is their ability to target specific immune cells, such as macrophages and dendritic cells, which play crucial roles in immune responses. Surface functionalization of NPs with specific ligands, such as chitosan and mannose, has been shown to enhance their uptake by these immune cells via specific phagocytic pathways. This targeted delivery to immune cells can improve the immune response to NP-bound antigens, such as those used in vaccine carriers [68]. In the context of cancer immunotherapy, nanotherapeutics have been investigated for their potential to modulate immunosuppressive cells and pathways, thereby potentiating the effectiveness of immunotherapy [87]. By targeting immunosuppressive cells, such as M2-like tumor-associated macrophages, and reducing the expression of immunosuppressive molecules, nanotherapeutics can enhance antitumor immune responses [88]. Besides, nanotherapeutics can display ligands that stimulate antitumor immune signals by engaging with immune cells and/or cancer cells [89]. These strategies aim to normalize the tumor immune microenvironment and broaden the impact of cancer immunotherapy [87]. Furthermore, nanotherapeutics can be designed to promote antitumor immune responses through immunopotentiating functions [89]. For example, they can target cancer cells to elicit immunogenic cell death, which can stimulate antitumor immune responses [88]. Nanotherapeutics can also target antigen-presenting cells, such as dendritic cells, to initiate immune activation and enhance antitumor immunity [90]. By modulating antitumor immune responses, nanotherapeutics have the potential to enhance the efficacy of cancer treatments and overcome biological barriers.

In terms of the mechanisms and signaling pathways involved in the recognition of nanotherapeutics by the immune system, there is still limited understanding. Most NPs do not possess common epitopes as expressed by other foreign antigens, and thus their recognition and processing by the immune system may differ [90]. Therefore, it is very important to elucidate these mechanisms and improve our ability to predict and prevent adverse immune reactions to nanotherapeutics. Nanotherapeutics for immune cell signaling have emerged as a promising approach in cancer immunotherapy. These nanomedicines are designed to enhance antitumor immune responses by promoting immunopotentiating functions. They can be directly injected into tumors or administered intravenously to achieve selective tumor accumulation. Tumor-targeted nanomedicines have the potential to enhance antitumor immunity through various mechanisms [89]. One approach is to trigger innate immune effector mechanisms by engaging with immune cells in tumors, such as tumor-associated macrophages (TAMs), natural killer (NK) cells, myeloid-derived suppressor cells (MDSCs), neutrophils, or dendritic cells (DCs) [89]. Nanotherapeutics can stimulate innate immune responses by delivering molecules that function as pathogen-associated molecular patterns (PAMPs) or danger-associated molecular patterns (DAMPs) [89]. This engagement with immune cells can enhance antitumor immune responses and contribute to the overall efficacy of nanotherapeutics. Another strategy is to modulate the tumor immune microenvironment. Nanotherapeutics can inhibit immunosuppressive cells, such as M2-like TAMs, and reduce the expression of immunosuppressive molecules, such as transforming growth factor beta [88]. By targeting the tumor immune microenvironment, nanomedicines can potentiate cancer immunotherapy and overcome immunosuppressive barriers. Furthermore, nanotherapeutics can target the peripheral immune system to potentiate antigen presentation in secondary lymphoid organs and activate peripheral immune cells [91]. This approach aims to reinforce the cancer-immunity cycle by promoting antigen release from cancer cells, antigen uptake and processing by APCs, and the presentation of cancer antigens to T cells [91]. By targeting the peripheral immune system, nanomedicines can stimulate T cells to recognize and kill cancer cells. In addition to their immunomodulatory functions, nanomedicines can also improve the delivery and biodistribution of immunotherapeutics. Biomimetic NPs, derived from cell membranes, have been developed to enhance the clinical efficacy of cancer immunotherapy [92]. These biomimetic nanotherapeutics have advantages such as improved efficacy compared to traditional therapies [92]. They can alter the landscape of cancer immunotherapy by improving the delivery and biodistribution of immunotherapeutics.

The interaction of NPs with the immune system is a crucial aspect to consider in the development of nanotherapeutics for immune cell signaling. NPs can be taken up by immune cells in the bloodstream and in tissues, including monocytes, platelets, leukocytes, DCs, macrophages, and B cells [68]. Understanding the interaction between NPs and immune cells is important for predicting NP biodistribution and potential effects on the immune system.

In short, NPs can target receptors and cell signaling pathways in immune cells through various mechanisms. The physicochemical properties of NPs, such as size and charge, can influence their uptake by immune cells. Surface functionalization with specific ligands or camouflaging with cellular membranes from immune cells can enhance NP uptake and immune response. The interaction between NPs and immune cells involves the activation of specific signaling pathways, which can further modulate the immune response. Understanding these interactions is crucial for the development of NP-based therapies and vaccines. Further study is desired to fully understand the mechanisms underlying the interaction between NPs and immune cells and to optimize the use of NPs in immunotherapy and vaccine development. By targeting specific immune cells and modulating immunosuppressive cells and pathways, nanotherapeutics can enhance antitumor immune responses and improve the efficacy of cancer immunotherapy. Further study is needed to optimize the design and efficacy of nanotherapeutics and to understand their interaction with the immune system.

4. Applications of NPs in immunotherapy

NP-based immunotherapy has emerged as a promising approach for the treatment of various diseases, including infectious diseases, cancer, and autoimmune disorders. The unique properties of NPs allow for targeted delivery of therapeutic agents, modulation of the immune response, and enhancement of the efficacy of immunotherapies [93–101].

4.1 Infectious Diseases

In the field of infectious diseases, NPs have shown potential as vaccine adjuvants and delivery systems [42]. They can enhance the immune response to vaccines by promoting antigen presentation and activation of immune cells. Gold NPs, for example, have been used to stimulate the activation of CD8+ T cells, leading to specific cytotoxic T-lymphocyte responses [42]. NPs can also be functionalized with antigens or immunomodulatory molecules to target specific pathogens and enhance the immune response against them [41].

4.2 Autoimmune diseases

Autoimmune diseases are characterized by an abnormal immune response against self-antigens, leading to chronic inflammation and tissue damage. NP-based treatments have shown promise in modulating the immune response and inducing immunological tolerance, which can be beneficial for the treatment of autoimmune diseases [73,102–105]. One approach in NP-based immunotherapy for autoimmune diseases is the induction of antigen-specific immunological tolerance. NPs can be designed to deliver antigens in a tolerogenic manner, promoting the development of regulatory T cells (Tregs) and dampening the immune response against self-antigens. These NPs can provide antigens alone or in combination with immunomodulatory molecules to harness natural tolerogenic processes and environments. For example, NPs can deliver antigens in the absence of costimulatory signals, mimic the tolerogenic environment of the liver, or promote apoptotic cell death to induce tolerance [73]. Furthermore, NPs can be used to target and expand autoregulatory T cells, such as memory-like autoregulatory CD8+ T cells, which can suppress autoreactivity and maintain immune homeostasis. NPs coated with specific antigens can stimulate the expansion of these autoregulatory T cells, providing a potential therapeutic strategy for autoimmune diseases [103]. In addition to antigen-specific tolerance induction, NP-based immunotherapy for autoimmune diseases can also focus on enhancing the function and number of Tregs, which are crucial for maintaining immune tolerance. NPs can be used to deliver autoantigens or immunomodulatory molecules to promote the expansion and activation of Tregs, restore immune homeostasis, and controll autoimmune responses [102,105]. The gut environment and gut microbiota have also been implicated in the pathogenesis of autoimmune diseases. NPs can modulate the gut microbiota and their metabolites, which play a critical role in immune modulation [106]. Short fatty acid chains, produced by the gut microbiota, have been shown to have a dual role in the pathogenesis of autoimmune disease models, and NPs can potentially target these metabolites to modulate immune responses [106].

4.3 Organ Transplantation

Organ transplantation is a life-saving procedure for patients with end-stage organ failure. However, the success of organ transplantation is limited by the risk of graft rejection and the need for long-term immunosuppression. In recent years, nanotechnology has emerged as a promising approach to improve the outcomes of organ transplantation by enhancing immune tolerance and reducing the need for immunosuppressive drugs [107–109]. One application of nanotherapeutics in organ transplantation is the modulation of immune responses to prevent graft rejection. NPs can be engineered to deliver immunosuppressive drugs directly to the transplanted organ, allowing for localized and sustained drug release [109,110]. For example, a study using a trained immunity-inhibiting and mechanistic target of rapamycin (mTOR)-specific nanotherapeutic showed prolonged allograft survival without the need for long-term immunosuppression in a mouse heart transplantation model [108]. This targeted delivery of immunosuppressive drugs can reduce systemic side effects and improve the efficacy of immunosuppression.

Nanotherapeutics can also be used to enhance the survival and function of transplanted cells. In the context of islet transplantation for type 1 diabetes, nanotechnology has been applied to improve the clinical potential of cell-based therapies [111]. NPs can be used to modify the surface of transplanted cells, enhancing their survival and protecting them from immune attack [111]. Additionally, NPs can be used to deliver prosurvival biomolecules to transplanted cells, improving their viability and function [112]. These approaches have the potential to improve the outcomes of cell-based therapies in organ transplantation. Furthermore, nanotherapeutics can be used to target APCs involved in transplant rejection. By specifically targeting APCs, nanotherapeutics can inhibit the upstream steps of transplant rejection and promote the generation of a durable donor-specific tolerant state [113]. This targeted approach can help prevent graft rejection and reduce the need for immunosuppressive drugs [113,114].

In addition to their immunomodulatory effects, nanotherapeutics can also be used for organ preservation and preconditioning. NPs can be used to deliver therapeutic agents to the organ during *ex vivo* normothermic machine perfusion (NMP), a technique used to preserve and assess the quality of donor organs prior to transplantation. This localized and controlled delivery of nanotherapeutics during NMP can optimize organ quality and reduce ischemia-reperfusion injury [110,115]. Moreover, nanotechnology has the potential to improve drug delivery in organ transplantation. NPs can be used as drug carriers to improve the pharmacokinetics and biodistribution of immunosuppressive drugs. By encapsulating drugs in NPs, their stability and solubility can be enhanced, leading to improved drug delivery to the target site. This targeted drug delivery can reduce systemic side effects and improve the efficacy of immunosuppression [116,117].

4.4   Cancer

Cancer immunotherapy is a promising approach that aims to utilize the power and specificity of the immune system to treat malignancies. Nanotechnology has emerged as a valuable tool in cancer immunotherapy, offering tremendous potential to enhance its effectiveness [118,119]. Recent advances in applying nanotechnologies to cancer immunotherapy have shown promising results. Nanotherapeutics have been developed to improve the delivery of immunotherapeutic agents and enhance immune responses [120]. These nanotherapeutics can significantly alter the landscape of cancer immunotherapy and improve its efficacy compared to traditional therapies [92]. One area of focus in cancer immunotherapy is the development of cancer vaccines. NP-based vaccines, also known as nanovaccines, have shown promise in stimulating immune responses against cancer. These nanovaccines can co-deliver antigens and adjuvants, enhancing the immune response and potentially providing a long-lasting vaccine effect [121]. NPs can also be used to target specific immune cells, such as lymphocytes or antigen-presenting cells, in circulating blood or lymphoid tissues, further enhancing the effectiveness of cancer immunotherapy [122].

In addition to cancer vaccines, nanotechnology can be used to remodel the tumor microenvironment and enhance immunotherapy [120]. NPs can be designed to modulate the tumor microenvironment, promoting immune cell infiltration and activation [123]. This can help overcome immunosuppressive factors in the tumor microenvironment and enhance the efficacy of immunotherapy [88]. Furthermore, nanotherapeutics can be used to deliver immunostimulatory agents and immune checkpoint inhibitors to the tumor site, improving their therapeutic effect [124]. Metallic nanotherapeutics have also shown promise in cancer immunotherapy. These NPs demonstrate promising cancer therapeutic and diagnostic efficiencies due to their surface functionalization ability, size distribution, and shape-dependent optical responses. They can be used for photothermal therapy and sonodynamic therapy, leading to apoptosis of cancer cells [125]. pH-responsive NPs have also been developed to provide spatiotemporal control of immunotherapeutic drugs, enhancing antitumor immunity and reducing off-tumor immunity [126]. Biomimetic nanotherapeutics, derived from cell membranes, have emerged as a novel approach in cancer immunotherapy. These nanotherapeutics mimic the properties of natural cells and can interact with the immune system in a more targeted and effective manner [92]. They have shown improved efficacy compared to traditional therapies and can potentially reshape the landscape of cancer immunotherapy.

The use of nanotechnology in cancer immunotherapy is not limited to traditional therapies. It also extends to emerging approaches such as adoptive cell therapy and therapeutic cancer vaccines [45]. NPs can be used to deliver therapeutic agents to immune cells, enhancing their function and improving the effectiveness of these therapies [127]. Furthermore, nanotechnology can be used to improve the delivery of immunostimulatory agents and monoclonal antibodies, providing new opportunities for cancer immunotherapy.

5.   Immunomodulatory nanotherapeutics

Immunomodulatory nanotherapeutics have gained significant attention in various fields of medicine, including regenerative medicine, cancer therapy, viral infections, age-related disorders, and skeletal muscle diseases. These nanotherapeutics offer the potential to modulate the immune system and enhance therapeutic outcomes. For example, cancer nanotherapeutics, which are designed to modulate the immune system's response to cancer cells, have shown improved efficacy compared to traditional therapies [92]. They can significantly alter the landscape of cancer immunotherapy by enhancing antitumor immunity through various mechanisms [89]. One approach to immunomodulatory nanotherapeutics is the use of biomimetic nanotechnology. Biomimetic nanotherapeutics are derived from cell membranes and can deliver immunomodulators to cancer cells, alter the tumor immune microenvironment, and stimulate innate immunity mechanisms [92]. These nanotherapeutics can induce immunogenic damage to cancer cells, eliminate or reprogram immunosuppressive cells, and enhance the exposure of tumor-associated antigens to antigen-presenting cells. By facilitating the infiltration of antitumor immune cells and their interaction with cancer cells, biomimetic nanotherapeutics can enhance antitumor immunity [89]. Nanomaterials have also been designed as immunomodulatory nanomedicines for cancer theranostics. These nanomaterials can be responsive to the tumor microenvironment and deliver immunomodulators to enhance the immune response against cancer cells. They have been shown to improve the efficacy of delivery systems and have the potential to advance to preclinical and clinical trials [128].

Tumor-targeted nanomedicines are another type of immunomodulatory nanotherapeutics that have shown promise in enhancing antitumor immunity. These nanomedicines can alter the tumor immune microenvironment by delivering immunomodulators, inducing immunogenic cell death in cancer cells, and reprogramming immunosuppressive cells [89]. They can also stimulate innate immunity mechanisms and facilitate the infiltration of antitumor immune cells into the tumor [89]. Further, NPs can be used to develop tumor-targeting T cells and CAR-T cell treatments, particularly for the treatment of solid tumors. These NPs can improve the delivery of immunotherapeutic agents to the tumor site and enhance the interaction between immune cells and cancer cells [129]. They can also be used to reshape the tumor microenvironment, creating a therapeutic window for effective T-cell therapy in solid malignancies [130]. In addition, nanodevices displaying tryptophan have been developed as immunomodulatory nanotherapeutics for treating acute lung injury. These nanodevices can modulate macrophage activation and cytokine production, leading to the prevention of acute lung injury [131]. NPs have also been used to inhibit inflammatory cytokines in lymphocytes and macrophages, providing a potential therapeutic approach for coronary artery disease [132].

6.   Challenges and future directions

6.1   Hurdles in translating NP-based approaches to clinical settings

Limitations in translating NP-based approaches to clinical settings arise from various factors. One major limitation is the heterogeneity of biological barriers across patient populations and diseases. NPs have been developed to overcome these barriers, including systemic, microenvironmental, and cellular barriers [133]. However, the effectiveness of NP-based drug delivery systems in improving overall patient survival compared to free drugs is limited [134]. This limitation highlights the need for a deeper understanding of clearance pathways and tumor targeting of NPs. Efforts from multiple disciplines are required to address these challenges and facilitate the translation of NP-based approaches into clinical practice.

Biodistribution and biological barrier breaching are among the biological issues that have limited the translation of NPs [133]. Additionally, the treatment of heterogeneous diseases poses a challenge for NP-based approaches. Technological issues, such as scale-up limitations and parameter optimization, also contribute to the limitations in clinical translation. Predicting the efficacy of NP-based approaches is another challenge that needs to be addressed. The low delivery efficiency of NPs to tumors is a critical barrier in their clinical translation. Although some improvements have been observed in delivery efficiencies, they are still relatively low. This limitation hinders the successful translation of nanomedicines into clinical practice [135]. Furthermore, the evaluation of surface-modified NP biocompatibility and the selection of appropriate in vivo models are current limitations in the field [136]. Standardized pathways for evaluating these aspects can facilitate the clinical translation of NP-based approaches.

The stability, circulation time, access, bioavailability, and safety profile of NPs are crucial factors for their successful clinical translation [137]. The clinical use of NPs has not always translated into improved clinical outcomes, despite their value in reducing drug toxicity [138]. Overcoming these limitations requires addressing challenges related to manufacturing, regulatory perspectives, and commercial development [139]. Reliable technologies with controlled manufacturing quality are essential for guaranteeing the clinical success of NP-based therapies. The limited clinical translation of NP-based approaches can be attributed to biological, immunological, and translational barriers [140]. Inorganic NPs offer opportunities to address unmet challenges in clinical settings. However, an incomplete understanding of nano-bio interactions has hindered the successful translation of NP-based therapeutic and imaging agents. The rapid clearance of NPs by the immune system poses an obstacle to their clinical translation.

The successful clinical translation of NP-based approaches is also hindered by limitations in NP toxicity and safety profiles. The development of safe-by-design approaches is necessary to ensure the clinical success of NP-based therapies [137]. Additionally, the complex synthesis and post-modification procedures associated with NP formulation pose challenges for reproducibly manufacturing clinical quality products [141]. Tailoring treatment regimens to individual patients is crucial for maximizing efficacy and minimizing toxicity.

Other limitations in the clinical translation of NP-based approaches include the limited tumor selectivity of NP-based materials [142]. Slow clearance of NPs and ambiguous biotoxicity evaluation hinder their bio-application in clinical settings [143]. The translation of NP-based systems from preclinical to clinical levels requires addressing barriers related to manufacturing, stability, and systemic toxicity evaluation [144]. The successful clinical translation of NP-based approaches also depends on an improved understanding of the enhanced permeability and retention effect [145].

The translation of NP-based approaches into clinical practice requires overcoming challenges related to implementation and integration into existing healthcare systems [146,147]. Structured framework-based approaches can facilitate the pragmatic and successful implementation of clinical trials. Additionally, partnerships and collaborations between researchers, clinicians, and community-based organizations are critical for integrating evidence-based prevention interventions within clinical and community settings. The competency-based approach and the

Nanomaterials in Biological Systems   Materials Research Forum LLC
Materials Research Foundations 185 (2026) 1-34   https://doi.org/10.21741/9781644903858-1

integration of research and clinical practice are essential for training and conducting psychological interventions in clinical settings.

In short, the translation of NP-based approaches to clinical settings is limited by various factors. These include the heterogeneity of biological barriers, limited improvements in overall patient survival compared to free drugs, challenges in clearance pathways and tumor targeting, biological and technological issues, low delivery efficiency, evaluation of biocompatibility and in vivo models, stability and safety profiles, manufacturing and regulatory perspectives, and implementation challenges. Overcoming these limitations requires interdisciplinary efforts, a deeper understanding of nano-bio interactions, safe-by-design approaches, improved manufacturing processes, tailored treatment regimens, and effective implementation strategies.

6.2   Future prospects for personalized immune modulation

Future prospects for personalized immune modulation hold great promise in the field of cancer immunotherapy. With the advancements in nanotechnology and our understanding of immune checkpoints, there are several potential avenues for personalized immune modulation.

One potential approach is the use of smart nanotherapeutics targeting tumor vasculature. These nanotherapeutics can specifically target unique biological molecules or tissues in the tumor vasculature, such as E-selectin, CD105, fibrin, tumor-associated pericytes, and endothelial cells [148]. By targeting these specific components, nanotherapeutics can increase tumor vessel permeability and improve the delivery of immune effector cells to the tumor site. Additionally, nanotherapeutics can correct the abnormal features of tumor vasculature, making it more efficient for drug delivery [148]. This targeted approach can enhance the efficacy of personalized immunotherapy.

Another potential avenue for personalized immune modulation is the regulation of immune checkpoints. Immune checkpoints play a crucial role in regulating immune responses, and their dysregulation can contribute to tumor immune evasion [149]. By modulating immune checkpoints, either through immune cells, oncolytic viruses, epigenetics, or gut microbiota, it is possible to enhance the antitumor immune response [149]. Predictions have been made for future personalized cancer immunotherapy based on different checkpoint modulations. Furthermore, the integration of nanotherapeutics with immunogenic gut microbiota has shown promise in reinstating anticancer adaptive T-cell responses against tumor cells [149]. By implanting immunogenic gut microbiota into specific patients, live commensals can be formulated to enhance the efficacy of anti-checkpoint therapy. This personalized approach takes into account the individual's gut microbiota composition and aims to strengthen the anticancer immune response.

In addition to these approaches, nanotherapeutics can also be used to deliver multiple therapeutic agents simultaneously, allowing for combined therapy and precise synchronization of antitumor attacks [148]. Nanotherapeutics can also integrate therapeutic drugs with molecular imaging agents, enabling theranostics in tumor vessels [148]. These strategies offer personalized treatment options by tailoring the therapy to the specific needs of each patient.

Thus, personalized immune modulation in cancer immunotherapy holds great potential for improving treatment outcomes. By utilizing nanotherapeutics, targeting tumor vasculature, regulating immune checkpoints, and integrating immunogenic gut microbiota, it is possible to enhance the efficacy and safety of immunotherapy. These approaches offer personalized treatment options that take into account the unique characteristics of each patient's tumor and immune system. However, it is important to note that further research is needed to optimize these approaches and translate them into clinical practice. The development of personalized immune modulation strategies requires a deep understanding of the underlying mechanisms and careful consideration of the individual patient's characteristics. With advancements in technology and our understanding of the immune system, personalized immune modulation has the potential to revolutionize cancer immunotherapy and improve patient outcomes.

Conclusion

NP-based approaches have shown great potential in the field of immunology, particularly in the context of cancer immunotherapy. These approaches enable combination treatment strategies that can make tumors with low immunogenicity susceptible to immunotherapy. NP-based therapeutic, prevention, and detection modalities have the potential to greatly impact how diseases are diagnosed and managed in the clinic.

NPs have been proposed as a potential polypharmacological strategy to modify multiple aspects of dysregulated immune responses, particularly in the context of severe inflammation and sepsis. They offer the potential to simultaneously modify multiple targets where single-target therapies have fallen short. NP-based therapeutics have the ability to overcome biological barriers, effectively deliver hydrophobic drugs and biologics, and preferentially target sites of disease.

In the field of immunology, NPs are believed to have strong potential for the development of advanced adjuvants, cytokines, vaccines, drugs, immunotherapies, and theranostic applications. Gold NPs, in particular, have been extensively studied and functionalized in various fields, including cellular and molecular biology, microbiology, immunology, and physiology. NPs can also be used for immune cytokine-based cancer therapy, such as tumor necrosis factor-related apoptosis-inducing ligand (TRAIL). Different platforms have been developed to deliver TRAIL using NPs, providing valuable insights for the design of future NP-based TRAIL cancer therapeutics.

Metal-based NPs have been shown to trigger innate and adaptive immune responses in various immune system models. The immunological activity of NPs is dependent on their physiochemical properties and subsequent cellular internalization. NPs have also been explored in the field of vaccination, including the development of NP-based vaccines for infectious diseases. Biocompatible and biodegradable polymeric NPs have gained interest as antigen delivery systems to enhance vaccine efficacy. Messenger RNA (mRNA) NPs have also been investigated as a potential vaccine delivery system, with advances in understanding mRNA's biological stability and immunological properties. Furthermore, NPs have been used to modulate immune responses in the context of specific diseases. For example, in the field of keloids, understanding the immunological mechanisms has led to the exploration of new therapeutic approaches.

In general, NP-based approaches have shown promise in various aspects of immunology, including cancer immunotherapy, vaccine development, modulation of immune responses, and targeted drug delivery. Continued research and exploration in this field are crucial to fully harness the potential of NP-based approaches in immunology.

Materials Research Forum LLC
https://doi.org/10.21741/9781644903858-1

References

[1] N. Subramanian, P. Torabi-Parizi, R.A. Gottschalk, R.N. Germain, B. Dutta, Network representations of immune system complexity, WIREs Syst. Biol. Med. 7 (2015) 13–38. https://doi.org/10.1002/wsbm.1288

[2] A.J. Husband, The immune system and integrated homeostasis, Immunol. Cell Biol. 73 (1995) 377–382. https://doi.org/10.1038/icb.1995.58

[3] N. Danilova, The evolution of immune mechanisms, J. Exp. Zoolog. B Mol. Dev. Evol. 306B (2006) 496–520. https://doi.org/10.1002/jez.b.21102

[4] D. Boraschi, P. Italiani, R. Palomba, P. Decuzzi, A. Duschl, B. Fadeel, S.M. Moghimi, Nanoparticles and innate immunity: new perspectives on host defence, Semin. Immunol. 34 (2017) 33–51. https://doi.org/10.1016/j.smim.2017.08.013

[5] X. Feng, W. Xu, Z. Li, W. Song, J. Ding, X. Chen, Immunomodulatory Nanosystems, Adv. Sci. 6 (2019) 1900101. https://doi.org/10.1002/advs.201900101

[6] B.D. Chithrani, A.A. Ghazani, W.C.W. Chan, Determining the size and shape dependence of gold nanoparticle uptake into mammalian cells, Nano Lett. 6 (2006) 662–668. https://doi.org/10.1021/nl052396o

[7] M.I. Montañez, A.J. Ruiz-Sanchez, E. Perez-Inestrosa, A perspective of nanotechnology in hypersensitivity reactions including drug allergy, Curr. Opin. Allergy Clin. Immunol. 10 (2010) 297–302. https://doi.org/10.1097/ACI.0b013e32833b1f17

[8] N. Yu, M. Ding, J. Li, Near-Infrared Photoactivatable Immunomodulatory Nanoparticles for Combinational Immunotherapy of Cancer, Front. Chem. 9 (2021). https://www.frontiersin.org/articles/10.3389/fchem.2021.701427 (accessed October 24, 2023)

[9] C.M. Lappas, The immunomodulatory effects of titanium dioxide and silver nanoparticles, Food Chem. Toxicol. 85 (2015) 78–83

[10] Y.-X. Lin, Y. Wang, S. Blake, M. Yu, L. Mei, H. Wang, J. Shi, RNA Nanotechnology-Mediated Cancer Immunotherapy, Theranostics 10 (2020) 281–299. https://doi.org/10.7150/thno.35568

[11] L. Yh, C. Lw, L. P, Metal-Based Nanoparticles and the Immune System: Activation, Inflammation, and Potential Applications, BioMed Res. Int. 2015 (2015). https://doi.org/10.1155/2015/143720

[12] D. La, Gold nanoparticles for preparation of antibodies and vaccines against infectious diseases, Expert Rev. Vaccines 19 (2020). https://doi.org/10.1080/14760584.2020.1758070

[13] C. Yang, W. Wang, K. Zhu, W. Liu, Y. Luo, X. Yuan, J. Wang, T. Cheng, X. Zhang, Lithium chloride with immunomodulatory function for regulating titanium nanoparticle-stimulated inflammatory response and accelerating osteogenesis through suppression of MAPK signaling pathway, Int. J. Nanomedicine Volume 14 (2019) 7475–7488. https://doi.org/10.2147/IJN.S210834

[14] M. Amina, N.M. Al Musayeib, N.A. Alarfaj, M.F. El-Tohamy, G.A. Al-Hamoud, Antibacterial and immunomodulatory potentials of biosynthesized Ag, Au, Ag-Au bimetallic

alloy nanoparticles using the Asparagus racemosus root extract, Nanomaterials 10 (2020) 2453.

[15]   Z. Li, Y. Hu, Q. Fu, Y. Liu, J. Wang, J. Song, H. Yang, NIR/ROS-Responsive Black Phosphorus QD Vesicles as Immunoadjuvant Carrier for Specific Cancer Photodynamic Immunotherapy, Adv. Funct. Mater. 30 (2020) 1905758. https://doi.org/10.1002/adfm.201905758

[16]   C.-M.J. Hu, R.H. Fang, B.T. Luk, K.N. Chen, C. Carpenter, W. Gao, K. Zhang, L. Zhang, 'Marker-of-self' functionalization of nanoscale particles through a top-down cellular membrane coating approach, Nanoscale 5 (2013) 2664–2668.

[17]   W.-H. Li, Y.-M. Li, Chemical Strategies to Boost Cancer Vaccines, Chem. Rev. 120 (2020) 11420–11478. https://doi.org/10.1021/acs.chemrev.9b00833

[18]   N.E. Papaioannou, O.V. Beniata, P. Vitsos, O. Tsitsilonis, P. Samara, Harnessing the immune system to improve cancer therapy, Ann. Transl. Med. 4 (2016) 261. https://doi.org/10.21037/atm.2016.04.01

[19]   C. Pardeshi, P. Rajput, V. Belgamwar, A. Tekade, G. Patil, K. Chaudhary, A. Sonje, Solid lipid based nanocarriers: an overview, Acta Pharm. Zagreb Croat. 62 (2012) 433–472. https://doi.org/10.2478/v10007-012-0040-z

[20]   A. Huppertsberg, L. Kaps, Z. Zhong, S. Schmitt, J. Stickdorn, K. Deswarte, F. Combes, C. Czysch, J. De Vrieze, S. Kasmi, N. Choteschovsky, A. Klefenz, C. Medina-Montano, P. Winterwerber, C. Chen, M. Bros, S. Lienenklaus, N.N. Sanders, K. Koynov, D. Schuppan, B.N. Lambrecht, S.A. David, B.G. De Geest, L. Nuhn, Squaric Ester-Based, pH-Degradable Nanogels: Modular Nanocarriers for Safe, Systemic Administration of Toll-like Receptor 7/8 Agonistic Immune Modulators, J. Am. Chem. Soc. 143 (2021) 9872–9883. https://doi.org/10.1021/jacs.1c03772

[21]   V. Malik, A. Ramesh, A.A. Kulkarni, TLR7/8 Agonist and SHP2 Inhibitor Loaded Nanoparticle Enhances Macrophage Immunotherapy Efficacy, Adv. Ther. 4 (2021) 2100086. https://doi.org/10.1002/adtp.202100086

[22]   Q.-Y. Ban, M. Liu, N. Ding, Y. Chen, Q. Lin, J.-M. Zha, W.-Q. He, Nutraceuticals for the Treatment of IBD: Current Progress and Future Directions, Front. Nutr. 9 (2022). https://www.frontiersin.org/articles/10.3389/fnut.2022.794169 (accessed October 24, 2023)

[23]   P. Sahdev, L.J. Ochyl, J.J. Moon, Biomaterials for Nanoparticle Vaccine Delivery Systems, Pharm. Res. 31 (2014) 2563–2582. https://doi.org/10.1007/s11095-014-1419-y

[24]   P. Italiani, G. Della Camera, D. Boraschi, Induction of Innate Immune Memory by Engineered Nanoparticles in Monocytes/Macrophages: From Hypothesis to Reality, Front. Immunol. 11 (2020). https://www.frontiersin.org/articles/10.3389/fimmu.2020.566309 (accessed October 24, 2023)

[25]   K. Chen, K. Xie, Z. Liu, Y. Nakasone, K. Sakao, M.A. Hossain, D.-X. Hou, Preventive effects and mechanisms of garlic on dyslipidemia and gut microbiome dysbiosis, Nutrients 11 (2019) 1225.

[26]   S. Muzammil, J. Neves Cruz, R. Mumtaz, I. Rasul, S. Hayat, M.A. Khan, A.M. Khan, M.U. Ijaz, R.R. Lima, M. Zubair, Effects of Drying Temperature and Solvents on In Vitro

Diabetic Wound Healing Potential of Moringa oleifera Leaf Extracts, Molecules 28 (2023) 710. https://doi.org/10.3390/molecules28020710

[27]   K. Inoue, H. Takano, R. Yanagisawa, S. Hirano, M. Sakurai, A. Shimada, T. Yoshikawa, Effects of Airway Exposure to Nanoparticles on Lung Inflammation Induced by Bacterial Endotoxin in Mice, Environ. Health Perspect. 114 (2006) 1325–1330. https://doi.org/10.1289/ehp.8903

[28]   I. Huizar, A. Malur, Y.A. Midgette, C. Kukoly, P. Chen, P.C. Ke, R. Podila, A.M. Rao, C.J. Wingard, L. Dobbs, B.P. Barna, M.S. Kavuru, M.J. Thomassen, Novel Murine Model of Chronic Granulomatous Lung Inflammation Elicited by Carbon Nanotubes, Am. J. Respir. Cell Mol. Biol. 45 (2011) 858–866. https://doi.org/10.1165/rcmb.2010-0401OC

[29]   R. Huq, E.L.G. Samuel, W.K.A. Sikkema, L.G. Nilewski, T. Lee, M.R. Tanner, F.S. Khan, P.C. Porter, R.B. Tajhya, R.S. Patel, T. Inoue, R.G. Pautler, D.B. Corry, J.M. Tour, C. Beeton, Preferential uptake of antioxidant carbon nanoparticles by T lymphocytes for immunomodulation, Sci. Rep. 6 (2016) 33808. https://doi.org/10.1038/srep33808

[30]   B. Zhang, Y. Su, J. Zhou, Y. Zheng, D. Zhu, Toward a Better Regeneration through Implant-Mediated Immunomodulation: Harnessing the Immune Responses, Adv. Sci. 8 (2021) 2100446. https://doi.org/10.1002/advs.202100446

[31]   T.L. Nguyen, Y. Choi, J. Kim, Mesoporous Silica as a Versatile Platform for Cancer Immunotherapy, Adv. Mater. Deerfield Beach Fla 31 (2019) e1803953. https://doi.org/10.1002/adma.201803953

[32]   Y. Yang, Z. Gu, J. Tang, M. Zhang, Y. Yang, H. Song, C. Yu, MnO$_2$ Nanoflowers Induce Immunogenic Cell Death under Nutrient Deprivation: Enabling an Orchestrated Cancer Starvation-Immunotherapy, Adv. Sci. 8 (2021) 2002667. https://doi.org/10.1002/advs.202002667

[33]   M. Kroker, U. Sydlik, A. Autengruber, C. Cavelius, H. Weighardt, A. Kraegeloh, K. Unfried, Preventing carbon nanoparticle-induced lung inflammation reduces antigen-specific sensitization and subsequent allergic reactions in a mouse model, Part. Fibre Toxicol. 12 (2015) 20. https://doi.org/10.1186/s12989-015-0093-5

[34]   Q. Jiao, L. Li, Q. Mu, Q. Zhang, Immunomodulation of nanoparticles in nanomedicine applications, BioMed Res. Int. 2014 (2014) 426028. https://doi.org/10.1155/2014/426028

[35]   W. Huang, H. Jia, J. Luo, Study on Dynamics and Cluster Analysis between Carbon-based Nanoparticles and Amino Acids, (2021). https://www.preprints.org/manuscript/202111.0406 (accessed October 24, 2023)

[36]   W. Huang, Z. Wang, J. Luo, Molecular Dynamics Study of the Curvature-Driven Interactions between Carbon-Based Nanoparticles and Amino Acids, Molecules 28 (2023) 482. https://doi.org/10.3390/molecules28020482

[37]   S.O. Obende, C.O. Ochieng, E.A. Shikanga, J.N. Cruz, C.B.R. Santos, N.M. Kimani, Croton's therapeutic promise: A review of its phytochemistry and critical computational ADME/Tox analysis, South Afr. J. Bot. 171 (2024) 648–672. https://doi.org/10.1016/j.sajb.2024.06.031

[38]   X. Qu, D. Zhou, J. Lu, D. Qin, J. Zhou, H.-J. Liu, Cancer nanomedicine in preoperative therapeutics: Nanotechnology-enabled neoadjuvant chemotherapy, radiotherapy, immunotherapy, and phototherapy, Bioact. Mater. 24 (2023) 136–152. https://doi.org/10.1016/j.bioactmat.2022.12.010

[39]   P.-Y. Xu, X. Zheng, R.K. Kankala, S.-B. Wang, A.-Z. Chen, Advances in Indocyanine Green-Based Codelivery Nanoplatforms for Combinatorial Therapy, ACS Biomater. Sci. Eng. 7 (2021) 939–962. https://doi.org/10.1021/acsbiomaterials.0c01644

[40]   A. Sentkowska, K. Pyrzynska, Does the Type Matter? Verification of Different Tea Types' Potential in the Synthesis of SeNPs, Antioxid. Basel Switz. 11 (2022) 2489. https://doi.org/10.3390/antiox11122489

[41]   M. Anwar, F. Muhammad, B. Akhtar, M.I. Anwar, A. Raza, A. Aleem, Outer Membrane Protein-Coated Nanoparticles as Antibacterial Vaccine Candidates, Int. J. Pept. Res. Ther. 27 (2021) 1689–1697. https://doi.org/10.1007/s10989-021-10201-3

[42]   L.A. Dykman, Gold nanoparticles for preparation of antibodies and vaccines against infectious diseases, Expert Rev. Vaccines 19 (2020) 465–477. https://doi.org/10.1080/14760584.2020.1758070

[43]   S. Kamat, M. Kumari, C. Jayabaskaran, Nano-engineered tools in the diagnosis, therapeutics, prevention, and mitigation of SARS-CoV-2, J. Controlled Release 338 (2021) 813–836. https://doi.org/10.1016/j.jconrel.2021.08.046

[44]   C. Oh, C. Lee, J. Park, J. Choi, S. Lee, S. Li, H.N. Jung, J.-S. Lee, J.-E. Hwang, J. Park, M. Kim, S. Baek, H.-J. Im, Development of Spleen Targeting H2S Donor Loaded Liposome for the Effective Systemic Immunomodulation and Treatment of Inflammatory Bowel Disease, ACS Nano 17 (2023) 4327–4345. https://doi.org/10.1021/acsnano.2c08898

[45]   H. Yan, D. Shao, Y. Lao, M. Li, H. Hu, K.W. Leong, Engineering Cell Membrane-Based Nanotherapeutics to Target Inflammation, Adv. Sci. 6 (2019) 1900605. https://doi.org/10.1002/advs.201900605

[46]   I.N. de F. Ramos, M.F. da Silva, J.M.S. Lopes, J.N. Cruz, F.S. Alves, J. de A.R. do Rego, M.L. da Costa, P.P. de Assumpção, D. do S. Barros Brasil, A.S. Khayat, Extraction, Characterization, and Evaluation of the Cytotoxic Activity of Piperine in Its Isolated form and in Combination with Chemotherapeutics against Gastric Cancer, Molecules 28 (2023) 5587. https://doi.org/10.3390/molecules28145587

[47]   A.C. Reisetter, L.V. Stebounova, J. Baltrusaitis, L. Powers, A. Gupta, V.H. Grassian, M.M. Monick, Induction of Inflammasome-dependent Pyroptosis by Carbon Black Nanoparticles *, J. Biol. Chem. 286 (2011) 21844–21852. https://doi.org/10.1074/jbc.M111.238519

[48]   G. Ibrahim Fouad, A proposed insight into the anti-viral potential of metallic nanoparticles against novel coronavirus disease-19 (COVID-19), Bull. Natl. Res. Cent. 45 (2021) 36. https://doi.org/10.1186/s42269-021-00487-0

[49]   W. Park, Y.-J. Heo, D.K. Han, New opportunities for nanoparticles in cancer immunotherapy, Biomater. Res. 22 (2018) 24. https://doi.org/10.1186/s40824-018-0133-y

[50]   W. Hou, B. Yang, H. Zhu, Nanoparticle-Based Therapeutic Strategies for Enhanced Pancreatic Ductal Adenocarcinoma Immunotherapy, Pharmaceutics 14 (2022) 2033.

[51]   A. Aufschnaiter, V. Kohler, S. Khalifa, A. Abd El-Wahed, M. Du, H. El-Seedi, S. Büttner, Apitoxin and its components against cancer, neurodegeneration and rheumatoid arthritis: Limitations and possibilities, Toxins 12 (2020) 66.

[52]   J. Kim, A. Narayana, S. Patel, G. Sahay, Advances in intracellular delivery through supramolecular self-assembly of oligonucleotides and peptides, Theranostics 9 (2019) 3191–3212. https://doi.org/10.7150/thno.33921

[53]   H. Qin, Y. Ding, A. Mujeeb, Y. Zhao, G. Nie, Tumor microenvironment targeting and responsive peptide-based nanoformulations for improved tumor therapy, Mol. Pharmacol. 92 (2017) 219–231.

[54]   L. Yang, K.D. Patel, C. Rathnam, R. Thangam, Y. Hou, H. Kang, K. Lee, Harnessing the Therapeutic Potential of Extracellular Vesicles for Biomedical Applications Using Multifunctional Magnetic Nanomaterials, Small 18 (2022) 2104783. https://doi.org/10.1002/smll.202104783

[55]   D. Murugan, V. Murugesan, B. Panchapakesan, L. Rangasamy, Nanoparticle enhancement of natural killer (NK) cell-based immunotherapy, Cancers 14 (2022) 5438.

[56]   M.L. Carmo Bastos, J.V. Silva-Silva, J. Neves Cruz, A.R. Palheta da Silva, A.A. Bentaberry-Rosa, G. da Costa Ramos, J.E. de Sousa Siqueira, M.R. Coelho-Ferreira, S. Percário, P. Santana Barbosa Marinho, A.M. do R. Marinho, M. de Oliveira Bahia, M.F. Dolabela, Alkaloid from Geissospermum sericeum Benth. & Hook.f. ex Miers (Apocynaceae) Induce Apoptosis by Caspase Pathway in Human Gastric Cancer Cells, Pharmaceuticals 16 (2023) 765. https://doi.org/10.3390/ph16050765

[57]   Y. Zeng, S. Li, S. Zhang, L. Wang, H. Yuan, F. Hu, Cell membrane coated-nanoparticles for cancer immunotherapy, Acta Pharm. Sin. B 12 (2022) 3233–3254. https://doi.org/10.1016/j.apsb.2022.02.023

[58]   U. Altanerova, M. Babincova, P. Babinec, K. Benejova, J. Jakubechova, V. Altanerova, M. Zduriencikova, V. Repiska, C. Altaner, Human mesenchymal stem cell-derived iron oxide exosomes allow targeted ablation of tumor cells via magnetic hyperthermia, Int. J. Nanomedicine Volume 12 (2017) 7923–7936. https://doi.org/10.2147/IJN.S145096

[59]   X. Michalet, F.F. Pinaud, L.A. Bentolila, J.M. Tsay, S. Doose, J.J. Li, G. Sundaresan, A.M. Wu, S.S. Gambhir, S. Weiss, Quantum Dots for Live Cells, in Vivo Imaging, and Diagnostics, Science 307 (2005) 538–544. https://doi.org/10.1126/science.1104274

[60]   C.E. Probst, P. Zrazhevskiy, V. Bagalkot, X. Gao, Quantum dots as a platform for nanoparticle drug delivery vehicle design, Adv. Drug Deliv. Rev. 65 (2013) 703–718. https://doi.org/10.1016/j.addr.2012.09.036

[61]   E. Przybytkowski, M. Behrendt, D. Dubois, D. Maysinger, Nanoparticles can induce changes in the intracellular metabolism of lipids without compromising cellular viability, FEBS J. 276 (2009) 6204–6217. https://doi.org/10.1111/j.1742-4658.2009.07324.x

[62]   J. Park, W. Gao, R. Whiston, T.B. Strom, S. Metcalfe, T.M. Fahmy, Modulation of CD4+ T Lymphocyte Lineage Outcomes with Targeted, Nanoparticle-Mediated Cytokine Delivery, Mol. Pharm. 8 (2011) 143–152. https://doi.org/10.1021/mp100203a

[63]   Z.R. Stephen, M. Zhang, Recent Progress in the Synergistic Combination of Nanoparticle-Mediated Hyperthermia and Immunotherapy for Treatment of Cancer, Adv. Healthc. Mater. 10 (2021) 2001415. https://doi.org/10.1002/adhm.202001415

[64]   F. Kavoosi, F. Modaresi, M. Sanaei, Z. Rezaei, Medical and dental applications of nanomedicines, APMIS 126 (2018) 795–803. https://doi.org/10.1111/apm.12890

[65]   M.H. Sarfraz, M. Zubair, B. Aslam, A. Ashraf, M.H. Siddique, S. Hayat, J.N. Cruz, S. Muzammil, M. Khurshid, M.F. Sarfraz, A. Hashem, T.M. Dawoud, G.D. Avila-Quezada, E.F. Abd_Allah, Comparative analysis of phyto-fabricated chitosan, copper oxide, and chitosan-based CuO nanoparticles: antibacterial potential against Acinetobacter baumannii isolates and anticancer activity against HepG2 cell lines, Front. Microbiol. 14 (2023) 1188743. https://doi.org/10.3389/fmicb.2023.1188743

[66]   Y. Zhao, X. Zhao, Y. Cheng, X. Guo, W. Yuan, Iron Oxide Nanoparticles-Based Vaccine Delivery for Cancer Treatment, Mol. Pharm. 15 (2018) 1791–1799. https://doi.org/10.1021/acs.molpharmaceut.7b01103

[67]   A.N. Ilinskaya, M.A. Dobrovolskaia, Immunosuppressive and anti-inflammatory properties of engineered nanomaterials, Br. J. Pharmacol. 171 (2014) 3988–4000. https://doi.org/10.1111/bph.12722

[68]   M.A. Dobrovolskaia, P. Aggarwal, J.B. Hall, S.E. McNeil, Preclinical Studies To Understand Nanoparticle Interaction with the Immune System and Its Potential Effects on Nanoparticle Biodistribution, Mol. Pharm. 5 (2008) 487–495. https://doi.org/10.1021/mp800032f

[69]   T.A. Ngobili, M.A. Daniele, Nanoparticles and direct immunosuppression, Exp. Biol. Med. 241 (2016) 1064–1073. https://doi.org/10.1177/1535370216650053

[70]   M. Di Gioacchino, L. Di Giampaolo, R. Mangifesta, S. Gangemi, C. Petrarca, Exposure to nanoparticles and occupational allergy, Curr. Opin. Allergy Clin. Immunol. 22 (2022) 55–63.

[71]   W. Yang, A.N. Frickenstein, V. Sheth, A. Holden, E.M. Mettenbrink, L. Wang, A.A. Woodward, B.S. Joo, S.K. Butterfield, N.D. Donahue, D.E. Green, A.G. Thomas, T. Harcourt, H. Young, M. Tang, Z.A. Malik, R.G. Harrison, P. Mukherjee, P.L. DeAngelis, S. Wilhelm, Controlling Nanoparticle Uptake in Innate Immune Cells with Heparosan Polysaccharides, Nano Lett. 22 (2022) 7119–7128. https://doi.org/10.1021/acs.nanolett.2c02226

[72]   X. Liang, X. Li, J. Duan, Y. Chen, X. Wang, L. Pang, D. Kong, B. Song, C. Li, J. Yang, Nanoparticles with CD44 Targeting and ROS Triggering Properties as Effective in Vivo Antigen Delivery System, Mol. Pharm. 15 (2018) 508–518. https://doi.org/10.1021/acs.molpharmaceut.7b00890

[73]   T.K. Kishimoto, R.A. Maldonado, Nanoparticles for the Induction of Antigen-Specific Immunological Tolerance, Front. Immunol. 9 (2018). https://www.frontiersin.org/articles/10.3389/fimmu.2018.00230 (accessed October 25, 2023)

[74]   Z. Deng, Y. Rong, Y. Teng, J. Mu, X. Zhuang, M. Tseng, A. Samykutty, L. Zhang, J. Yan, D. Miller, J. Suttles, H.-G. Zhang, Broccoli-Derived Nanoparticle Inhibits Mouse Colitis by Activating Dendritic Cell AMP-Activated Protein Kinase, Mol. Ther. 25 (2017) 1641–1654. https://doi.org/10.1016/j.ymthe.2017.01.025

[75]   D. Wang, T. Wang, H. Yu, B. Feng, L. Zhou, F. Zhou, B. Hou, H. Zhang, M. Luo, Y. Li, Engineering nanoparticles to locally activate T cells in the tumor microenvironment, Sci. Immunol. 4 (2019) eaau6584. https://doi.org/10.1126/sciimmunol.aau6584

[76]   V. Selvaraj, N. Nepal, S. Rogers, N.D. Manne, R. Arvapalli, K.M. Rice, S. Asano, E. Fankhanel, J.J. Ma, T. Shokuhfar, Inhibition of MAP kinase/NF-kB mediated signaling and attenuation of lipopolysaccharide induced severe sepsis by cerium oxide nanoparticles, Biomaterials 59 (2015) 160–171.

[77]   X. Hou, Y. Tao, Y. Pang, X. Li, G. Jiang, Y. Liu, Nanoparticle-based photothermal and photodynamic immunotherapy for tumor treatment, Int. J. Cancer 143 (2018) 3050–3060. https://doi.org/10.1002/ijc.31717

[78]   Q. Meng, R. Tian, H. Long, X. Wu, J. Lai, O. Zharkova, J. Wang, X. Chen, L. Rao, Capturing Cytokines with Advanced Materials: A Potential Strategy to Tackle COVID-19 Cytokine Storm, Adv. Mater. 33 (2021) 2100012. https://doi.org/10.1002/adma.202100012.

[79]   S.G. Tangye, K.L. Good, Human IgM+ CD27+ B cells: memory B cells or "memory" B cells?, J. Immunol. 179 (2007) 13–19.

[80]   A.G. Lenz, E. Karg, B. Lentner, V. Dittrich, C. Brandenberger, B. Rothen-Rutishauser, H. Schulz, G.A. Ferron, O. Schmid, A dose-controlled system for air-liquid interface cell exposure and application to zinc oxide nanoparticles, Part. Fibre Toxicol. 6 (2009) 32. https://doi.org/10.1186/1743-8977-6-32

[81]   Y. Zhu, J.W. Eaton, C. Li, Titanium Dioxide (TiO2) Nanoparticles Preferentially Induce Cell Death in Transformed Cells in a Bak/Bax-Independent Fashion, PLOS ONE 7 (2012) e50607. https://doi.org/10.1371/journal.pone.0050607

[82]   A.C. Anselmo, M. Zhang, S. Kumar, D.R. Vogus, S. Menegatti, M.E. Helgeson, S. Mitragotri, Elasticity of Nanoparticles Influences Their Blood Circulation, Phagocytosis, Endocytosis, and Targeting, ACS Nano 9 (2015) 3169–3177. https://doi.org/10.1021/acsnano.5b00147

[83]   M. Zhu, L. Du, R. Zhao, H.Y. Wang, Y. Zhao, G. Nie, R.-F. Wang, Cell-Penetrating Nanoparticles Activate the Inflammasome to Enhance Antibody Production by Targeting Microtubule-Associated Protein 1-Light Chain 3 for Degradation, ACS Nano 14 (2020) 3703–3717. https://doi.org/10.1021/acsnano.0c00962

[84]   J.F. Hernandez-Franco, S. Xie, J. Thimmapuram, D. Ragland, H. HogenEsch, Mechanism of activation of porcine dendritic cells by an α-D-glucan nanoparticle adjuvant and a nanoparticle/poly(I:C) combination adjuvant, Front. Immunol. 13 (2022). https://www.frontiersin.org/articles/10.3389/fimmu.2022.990900 (accessed October 25, 2023)

[85]   B. Carrillo-Conde, E.-H. Song, A. Chavez-Santoscoy, Y. Phanse, A.E. Ramer-Tait, N.L.B. Pohl, M.J. Wannemuehler, B.H. Bellaire, B. Narasimhan, Mannose-Functionalized "Pathogen-like" Polyanhydride Nanoparticles Target C-Type Lectin Receptors on Dendritic Cells, Mol. Pharm. 8 (2011) 1877–1886. https://doi.org/10.1021/mp200213r

[86]   V. Vijayan, S. Uthaman, I.-K. Park, Cell membrane-camouflaged nanoparticles: a promising biomimetic strategy for cancer theragnostics, Polymers 10 (2018) 983.

[87]   M. Saeed, F. Chen, J. Ye, Y. Shi, T. Lammers, B.G. De Geest, Z.P. Xu, H. Yu, From Design to Clinic: Engineered Nanobiomaterials for Immune Normalization Therapy of Cancer, Adv. Mater. 33 (2021) 2008094. https://doi.org/10.1002/adma.202008094

[88]   Y. Shi, T. Lammers, Combining Nanomedicine and Immunotherapy, Acc. Chem. Res. 52 (2019) 1543–1554. https://doi.org/10.1021/acs.accounts.9b00148.

[89]   H. Cabral, H. Kinoh, K. Kataoka, Tumor-Targeted Nanomedicine for Immunotherapy, Acc. Chem. Res. 53 (2020) 2765–2776. https://doi.org/10.1021/acs.accounts.0c00518

[90]   M. Zhu, X. Tian, X. Song, Y. Li, Y. Tian, Y. Zhao, G. Nie, Nanoparticle-Induced Exosomes Target Antigen-Presenting Cells to Initiate Th1-Type Immune Activation, Small 8 (2012) 2841–2848. https://doi.org/10.1002/smll.201200381

[91]   Q. Sun, X. Bai, A.M. Sofias, R. van der Meel, E. Ruiz-Hernandez, G. Storm, W.E. Hennink, B. De Geest, F. Kiessling, H. Yu, Cancer nanomedicine meets immunotherapy: opportunities and challenges, Acta Pharmacol. Sin. 41 (2020) 954–958.

[92]   F. Raza, H. Zafar, S. Zhang, Z. Kamal, J. Su, W. Yuan, Q. Mingfeng, Recent Advances in Cell Membrane-Derived Biomimetic Nanotechnology for Cancer Immunotherapy, Adv. Healthc. Mater. 10 (2021) 2002081. https://doi.org/10.1002/adhm.202002081

[93]   C. Gomez-Casado, A. Villaseñor, A. Rodriguez-Nogales, J.L. Bueno, D. Barber, M.M. Escribese, Understanding Platelets in Infectious and Allergic Lung Diseases, Int. J. Mol. Sci. 20 (2019) 1730. https://doi.org/10.3390/ijms20071730

[94]   V.K. Selvaraja, D.K. Gudipudi, Fundamentals to clinical application of nanoparticles in cancer immunotherapy and radiotherapy, Ecancermedicalscience 14 (2020) 1095. https://doi.org/10.3332/ecancer.2020.1095

[95]   Z. Wang, Z. Wang, B. Li, S. Wang, T. Chen, Z. Ye, Innate immune cells: a potential and promising cell population for treating osteosarcoma, Front. Immunol. 10 (2019) 1114.

[96]   J. Ozao-Choy, G. Ma, J. Kao, G.X. Wang, M. Meseck, M. Sung, M. Schwartz, C.M. Divino, P.-Y. Pan, S.-H. Chen, The novel role of tyrosine kinase inhibitor in the reversal of immune suppression and modulation of tumor microenvironment for immune-based cancer therapies, Cancer Res. 69 (2009) 2514–2522.

[97]   Z.J. Brown, T.F. Greten, B. Heinrich, Adjuvant Treatment of Hepatocellular Carcinoma: Prospect of Immunotherapy, Hepatology 70 (2019) 1437–1442. https://doi.org/10.1002/hep.30633

[98]   J. Cui, G. Yang, Z. Pan, Y. Zhao, X. Liang, W. Li, L. Cai, Hormetic response to low-dose radiation: focus on the immune system and its clinical implications, Int. J. Mol. Sci. 18 (2017) 280.

[99]   D.J. Pinato, N. Guerra, P. Fessas, R. Murphy, T. Mineo, F.A. Mauri, S.K. Mukherjee, M. Thursz, C.N. Wong, R. Sharma, Immune-based therapies for hepatocellular carcinoma, Oncogene 39 (2020) 3620–3637.

[100]   N. Benne, D. Ter Braake, A.J. Stoppelenburg, F. Broere, Nanoparticles for inducing antigen-specific T cell tolerance in autoimmune diseases, Front. Immunol. 13 (2022) 864403.

[101]   D. Murugan, V. Murugesan, B. Panchapakesan, L. Rangasamy, Nanoparticle enhancement of natural killer (NK) cell-based immunotherapy, Cancers 14 (2022) 5438.

[102]   D.A. Horwitz, S. Bickerton, A. La Cava, Strategies to use nanoparticles to generate CD4 and CD8 regulatory T cells for the treatment of SLE and other autoimmune diseases, Front. Immunol. 12 (2021) 681062.

[103]   S. Tsai, A. Shameli, J. Yamanouchi, X. Clemente-Casares, J. Wang, P. Serra, Y. Yang, Z. Medarova, A. Moore, P. Santamaria, Reversal of autoimmunity by boosting memory-like autoregulatory T cells, Immunity 32 (2010) 568–580.

[104]   M. Mizuno, D. Noto, N. Kaga, A. Chiba, S. Miyake, The dual role of short fatty acid chains in the pathogenesis of autoimmune disease models, PloS One 12 (2017) e0173032.

[105]   P.J. Eggenhuizen, B.H. Ng, J.D. Ooi, Treg enhancing therapies to treat autoimmune diseases, Int. J. Mol. Sci. 21 (2020) 7015.

[106]   M. Mizuno, D. Noto, N. Kaga, A. Chiba, S. Miyake, The dual role of short fatty acid chains in the pathogenesis of autoimmune disease models, PloS One 12 (2017) e0173032.

[107]   E. Tasciotti, F.J. Cabrera, M. Evangelopoulos, J.O. Martinez, U.R. Thekkedath, M. Kloc, R.M. Ghobrial, X.C. Li, A. Grattoni, M. Ferrari, The emerging role of nanotechnology in cell and organ transplantation, Transplantation 100 (2016) 1629.

[108]   M.G. Netea, J. Domínguez-Andrés, L.B. Barreiro, T. Chavakis, M. Divangahi, E. Fuchs, L.A. Joosten, J.W. van der Meer, M.M. Mhlanga, W.J. Mulder, Defining trained immunity and its role in health and disease, Nat. Rev. Immunol. 20 (2020) 375–388.

[109]   A. Dangi, S. Yu, X. Luo, Emerging approaches and technologies in transplantation: the potential game changers, Cell. Mol. Immunol. 16 (2019) 334–342.

[110]   B. Hussain, V. Kasinath, J.C. Madsen, J. Bromberg, S.G. Tullius, R. Abdi, Intra-Organ Delivery of Nanotherapeutics for Organ Transplantation, ACS Nano 15 (2021) 17124–17136. https://doi.org/10.1021/acsnano.1c04707

[111]   S.C. Wiggins, N.J. Abuid, K.M. Gattás-Asfura, S. Kar, C.L. Stabler, Nanotechnology Approaches to Modulate Immune Responses to Cell-based Therapies for Type 1 Diabetes, J. Diabetes Sci. Technol. 14 (2020) 212–225. https://doi.org/10.1177/1932296819871947

[112]   L. Ferreira, J.M. Karp, L. Nobre, R. Langer, New opportunities: the use of nanotechnologies to manipulate and track stem cells, Cell Stem Cell 3 (2008) 136–146.

[113]   J. Ochando, M.S. Braza, Nanoparticle-based modulation and monitoring of antigen-presenting cells in organ transplantation, Front. Immunol. 8 (2017) 1888.

[114]   K.A. Hlavaty, X. Luo, L.D. Shea, S.D. Miller, Cellular and molecular targeting for nanotherapeutics in transplantation tolerance, Clin. Immunol. 160 (2015) 14–23.

[115] E.R. Thompson, L. Bates, I.K. Ibrahim, A. Sewpaul, B. Stenberg, A. McNeill, R. Figueiredo, T. Girdlestone, G.C. Wilkins, L. Wang, Novel delivery of cellular therapy to reduce ischemia reperfusion injury in kidney transplantation, Am. J. Transplant. 21 (2021) 1402–1414.

[116] O.C. Farokhzad, R. Langer, Impact of Nanotechnology on Drug Delivery, ACS Nano 3 (2009) 16–20. https://doi.org/10.1021/nn900002m

[117] C.G. Yao, P.N. Martins, Nanotechnology applications in transplantation medicine, Transplantation 104 (2020) 682–693.

[118] P.H. Lizotte, A.M. Wen, M.R. Sheen, J. Fields, P. Rojanasopondist, N.F. Steinmetz, S. Fiering, In situ vaccination with cowpea mosaic virus nanoparticles suppresses metastatic cancer, Nat. Nanotechnol. 11 (2016) 295–303.

[119] J.N. Blattman, P.D. Greenberg, Cancer Immunotherapy: A Treatment for the Masses, Science 305 (2004) 200–205. https://doi.org/10.1126/science.1100369

[120] S. Musetti, L. Huang, Nanoparticle-Mediated Remodeling of the Tumor Microenvironment to Enhance Immunotherapy, ACS Nano 12 (2018) 11740–11755. https://doi.org/10.1021/acsnano.8b05893

[121] M. Luo, H. Wang, Z. Wang, H. Cai, Z. Lu, Y. Li, M. Du, G. Huang, C. Wang, X. Chen, A STING-activating nanovaccine for cancer immunotherapy, Nat. Nanotechnol. 12 (2017) 648–654.

[122] Y. Mi, C.T. Hagan, B.G. Vincent, A.Z. Wang, Emerging Nano-/Microapproaches for Cancer Immunotherapy, Adv. Sci. 6 (2019) 1801847. https://doi.org/10.1002/advs.201801847

[123] H. Shen, T. Sun, H.H. Hoang, J.S. Burchfield, G.F. Hamilton, E.A. Mittendorf, M. Ferrari, Enhancing cancer immunotherapy through nanotechnology-mediated tumor infiltration and activation of immune cells, in: Semin. Immunol., Elsevier, 2017: pp. 114–122. https://www.sciencedirect.com/science/article/pii/S1044532317300088 (accessed October 29, 2023)

[124] P. Velpurisiva, A. Gad, B. Piel, R. Jadia, P. Rai, Nanoparticle design strategies for effective cancer immunotherapy, J. Biomed. Syd. NSW 2 (2017) 64.

[125] A. Mohapatra, S. Uthaman, I.-K. Park, External and internal stimuli-responsive metallic nanotherapeutics for enhanced anticancer therapy, Front. Mol. Biosci. 7 (2021) 597634.

[126] Y. Yan, H. Ding, pH-responsive nanoparticles for cancer immunotherapy: a brief review, Nanomaterials 10 (2020) 1613.

[127] J. Xiang, L. Xu, H. Gong, W. Zhu, C. Wang, J. Xu, L. Feng, L. Cheng, R. Peng, Z. Liu, Antigen-Loaded Upconversion Nanoparticles for Dendritic Cell Stimulation, Tracking, and Vaccination in Dendritic Cell-Based Immunotherapy, ACS Nano 9 (2015) 6401–6411. https://doi.org/10.1021/acsnano.5b02014

[128] A. Khan, F. Dias, S. Neekhra, B. Singh, R. Srivastava, Designing and immunomodulating multiresponsive nanomaterial for cancer theranostics, Front. Chem. 8 (2021) 631351.

[129] C.M. Hartshorn, M.S. Bradbury, G.M. Lanza, A.E. Nel, J. Rao, A.Z. Wang, U.B. Wiesner, L. Yang, P. Grodzinski, Nanotechnology Strategies To Advance Outcomes in Clinical Cancer Care, ACS Nano 12 (2018) 24–43. https://doi.org/10.1021/acsnano.7b05108

[130] F. Zhang, S.B. Stephan, C.I. Ene, T.T. Smith, E.C. Holland, M.T. Stephan, Nanoparticles that reshape the tumor milieu create a therapeutic window for effective T-cell therapy in solid malignancies, Cancer Res. 78 (2018) 3718–3730.

[131] L. Sun, R. Wang, C. Wu, J. Gong, H. Ma, S.-Y. Fung, H. Yang, The modulatory activity of tryptophan displaying nanodevices on macrophage activation for preventing acute lung injury, Front. Immunol. 12 (2021) 750128.

[132] R. Shabbir, A. Raza, A. Liaquat, S.U. Shah, S. Saeed, U. Sarwar, M. Hamza, F. Chudhary, Z. Hussain, N.M. Butt, Nanoparticles as a novel tool to inhibit inflammatory cytokines in human lymphocytes and macrophages of coronary artery disease, J. Pharm. Sci. 111 (2022) 1509–1521.

[133] A.C. Anselmo, S. Mitragotri, Nanoparticles in the clinic: An update, Bioeng. Transl. Med. 4 (2019) e10143. https://doi.org/10.1002/btm2.10143

[134] M. Yu, J. Zheng, Clearance Pathways and Tumor Targeting of Imaging Nanoparticles, ACS Nano 9 (2015) 6655–6674. https://doi.org/10.1021/acsnano.5b01320

[135] Y.-H. Cheng, C. He, J.E. Riviere, N.A. Monteiro-Riviere, Z. Lin, Meta-Analysis of Nanoparticle Delivery to Tumors Using a Physiologically Based Pharmacokinetic Modeling and Simulation Approach, ACS Nano 14 (2020) 3075–3095. https://doi.org/10.1021/acsnano.9b08142

[136] J.W. Shreffler, J.E. Pullan, K.M. Dailey, S. Mallik, A.E. Brooks, Overcoming hurdles in nanoparticle clinical translation: The influence of experimental design and surface modification, Int. J. Mol. Sci. 20 (2019) 6056.

[137] J.A. Damasco, S. Ravi, J.D. Perez, D.E. Hagaman, M.P. Melancon, Understanding nanoparticle toxicity to direct a safe-by-design approach in cancer nanomedicine, Nanomaterials 10 (2020) 2186.

[138] Z. Cheng, A. Al Zaki, J.Z. Hui, V.R. Muzykantov, A. Tsourkas, Multifunctional Nanoparticles: Cost Versus Benefit of Adding Targeting and Imaging Capabilities, Science 338 (2012) 903–910. https://doi.org/10.1126/science.1226338

[139] H. Ragelle, F. Danhier, V. Préat, R. Langer, D.G. Anderson, Nanoparticle-based drug delivery systems: a commercial and regulatory outlook as the field matures, Expert Opin. Drug Deliv. 14 (2017) 851–864. https://doi.org/10.1080/17425247.2016.1244187.

[140] P. Grodzinski, M. Kircher, M. Goldberg, A. Gabizon, Integrating Nanotechnology into Cancer Care, ACS Nano 13 (2019) 7370–7376. https://doi.org/10.1021/acsnano.9b04266.

[141] I. Rana, J. Oh, J. Baig, J.H. Moon, S. Son, J. Nam, Nanocarriers for cancer nano-immunotherapy, Drug Deliv. Transl. Res. 13 (2023) 1936–1954. https://doi.org/10.1007/s13346-022-01241-3.

[142] M. Benezra, O. Penate-Medina, P.B. Zanzonico, D. Schaer, H. Ow, A. Burns, E. DeStanchina, V. Longo, E. Herz, S. Iyer, Multimodal silica nanoparticles are effective cancer-targeted probes in a model of human melanoma, J. Clin. Invest. 121 (2011) 2768–2780.

[143] X. Wang, L. Guo, S. Zhang, Y. Chen, Y.-T. Chen, B. Zheng, J. Sun, Y. Qian, Y. Chen, B. Yan, W. Lu, Copper Sulfide Facilitates Hepatobiliary Clearance of Gold Nanoparticles through the Copper-Transporting ATPase ATP7B, ACS Nano 13 (2019) 5720–5730. https://doi.org/10.1021/acsnano.9b01154.

[144] C. Zhang, Y.-Y. Fu, X. Zhang, C. Yu, Y. Zhao, S.-K. Sun, BSA-directed synthesis of CuS nanoparticles as a biocompatible photothermal agent for tumor ablation in vivo, Dalton Trans. 44 (2015) 13112–13118.

[145] V. Ejigah, O. Owoseni, P. Bataille-Backer, O.D. Ogundipe, F.A. Fisusi, S.K. Adesina, Approaches to improve macromolecule and nanoparticle accumulation in the tumor microenvironment by the enhanced permeability and retention effect, Polymers 14 (2022) 2601.

[146] L.A. Shrier, P.J. Burke, C. Jonestrask, S.L. Katz-Wise, Applying Systems Thinking and Human-Centered Design to Development of Intervention Implementation Strategies: An Example from Adolescent Health Research, J. Public Health Res. 9 (2020) jphr.2020.1746. https://doi.org/10.4081/jphr.2020.1746.

[147] P. Fathi, H.J. Knox, D. Sar, I. Tripathi, F. Ostadhossein, S.K. Misra, M.B. Esch, J. Chan, D. Pan, Biodegradable Biliverdin Nanoparticles for Efficient Photoacoustic Imaging, ACS Nano 13 (2019) 7690–7704. https://doi.org/10.1021/acsnano.9b01201.

[148] Z. Li, C. Di, S. Li, X. Yang, G. Nie, Smart Nanotherapeutic Targeting of Tumor Vasculature, Acc. Chem. Res. 52 (2019) 2703–2712. https://doi.org/10.1021/acs.accounts.9b00283.

[149] T. Shi, Y. Ma, L. Yu, J. Jiang, S. Shen, Y. Hou, T. Wang, Cancer immunotherapy: a focus on the regulation of immune checkpoints, Int. J. Mol. Sci. 19 (2018) 1389.

Nanomaterials in Biological Systems
Materials Research Foundations 185 (2026) 35-

Materials Research Forum LLC
https://doi.org/10.21741/9781644903858-2

Chapter 2

Studies of Nanoparticle Applications in Biology

M.V. Lakshmi[1], Sisanth K S[2], Remya Krishanan[3], Jiji Abraham[4], Sabu Thomas[2] ,T.S. Swapna[1*]

[1]Department of Botany, University of Kerala, Thiruvananthapuram, Kerala, India

[2]International and Inter University Centre for Nanoscience and Nanotechnology, Mahatma Gandhi University, Kottayam, India

[3]Department of Botany, Daulat Ram College, University of Delhi, Delhi, India

[4]Department of Chemistry, Vimala college, Thrissur, India

swapnats@yahoo.com

Abstract

Nanoparticles (NPs) encompass innumerable applications in biological sciences as they possess required physiochemical features without any detrimental effects. This chapter explained numerous valuable uses of NPs in biology. Due to their advancement in various practical approaches, pharmaceutical and other industries have got tremendous results. Methods of synthesis and modifications in their structure or properties influence their therapeutic potential. Hence, synthesis and characterization of NPs should give proper importance before finding any applications. NPs are extensively employed in distinct branches of biology with promising results such as imaging and detection, drug delivery, gene therapy, tissue engineering, cancer treatment and as antimicrobial agents and biosensors. Case studies were analysed for elaborating the diverse situations in which NPs find interest. The study also gives attention in defining the toxicological effects of NPs which may have detrimental effects in future directions.

Keywords

Nanomaterials, Characterization, Drug Delivery, Gene Therapy, Biosensors, Tissue Engineering, Biomedical applications

Contents

1.　Introduction to Nanoparticles and their Properties

Nanoparticles (NPs) are manually designed tiny particles of matter that are a few degrees larger than an atom, having variable sizes between 1 to 100nm. They have considerable recognition in nanotechnology research due to their outstanding optical, mechanical, electromagnetic, and optical properties [1]. There are different approaches for obtaining nanoparticles, including physical, chemical and biological methods. The practical application of these nanoparticles has made tremendous advances in agriculture, pharmaceutical, cosmetic, electronic and textile industries [2]. The large specific surface area and quantum effects cause NPs to behave differently than the same materials at larger dimensions [3]. Depending on the shape and size of these particles, their colour and properties are varied that are exploited in bioimaging applications [4]. NPs are usually seen in uniform or in several layers like surface layer, shell layer and the core layer [5]. Surface layer has huge diversity of small sized molecules, ionic metals, surfactants or polymers, and the shell layer consists of a chemically varied material from the central core layer; since the core layer is the central part of the NPs usually the Nanoparticles themself [6].

Researchers have reported various kinds of NPs such as organic, carbon-based, and inorganic nanomaterials on the physico-chemical features [7]. As their name depicts, organic NPs or polymers are found in nature that are composed of proteins, carbohydrates, lipids, polymers, or any other organic compounds and extensively utilized in many industrial products [8]. They may comprise solid/hard material together with organic components such as either polymers or lipids; specially at scales ranging from 10 nm to 1 m [9]. Polymer nanoparticles (PNPs) and lipid NPs, coming under the organic based NPs are effectively used in many biomedical applications [10]. Organic nanomaterials have versatile applications in molecular imaging, pharmaceutical formulations, and image-guided therapies [11].

Carbon-based NPs are the nanomaterials that have been studied in depth and produced in the various forms like the fullerenes, carbon nanotubes, graphene and also their derivatives like graphene oxide, nanodiamonds, and carbon-based quantum dots [12]. These are also important in various biomedical applications, including bio-sensing, drug delivery, and cancer therapy [13].

Inorganic nanoparticles can be cheaply synthesized and have been used in different fields like biomedicine, electronics, storage media, conservation of cultural heritage, optics, textiles, and cosmetics [14]. They include metallic as well as non-metallic nanoparticles. There are a variety of inorganic non-metallic nanoparticles that comprise of titania-based ceramics, alumina ceramics, calcium phosphate (CaP), tricalcium phosphate (TCP), hydroxyapatite (HAP), calcium sulfate and calcium carbonate, and bioactive glass ceramics [15]. Metallic gold, silver, and magnetic nanoparticles (iron oxide) have been employed as diagnostic and therapeutic agents on a large scale [16].

2. Nanoparticles Synthesis Methods

Numerous techniques have been developed for the synthesis of NPs. The basic physical, chemical and biological methods are usually followed for producing metallic nanoparticles. The methods such as microwaves, ultrasound, irradiation, or mechanical grinding, sonication [17]. Microgrinding, lithographic and wire explosion techniques are the very common physical techniques for NPs synthesis. In the physical method, the reducing or capping agents used are expensive [18] So, there is no problem of contamination of the final nanoparticles with solvent [19].

The nanoparticles synthesized through chemical methods can be stored for long periods of time without significant loss in stability [20]. It is a widely used economic method for the large-scale production of NPs. For example for synthesising chitosan nanoparticles (NP) - Polyelectrolyte complex (PEC) formulation is employed, the electrostatic interaction in between anions and cations succeeded by charge neutralization [21]. Another technique is the ionotropic gelation method where chitosan nanoparticles are engineered using an ionic cross-linked. Further-more, Microemulsion method, Covalent cross-linking method, Co-precipitation method, and Complex coacervation method can also be used for the synthesis of Chitosan nanoparticles [22]. But these methods, hazardous and highly toxic chemicals are used in this method for the reduction of various metals to nanoparticles. Here, there will be contamination of the nanoparticles with the hazardous byproducts[23].

The more sustainable and eco-friendly technique is the green synthesis of NPs when compared to chemical and physical methods. In this method, reducing agents are administered from plant extracts or microorganisms. However, it is not possible for all the metal nanoparticles to remove the environmental pollutants in biological methods [24]. Nanoparticles can be biologically synthesized with the aid of the microorganisms/mycelium, which target specific ions which are from their surroundings and then transform the ionic metals into the metal element through enzymes which are generated by the cell activities [25]. It can be categorized into various ways of intracellular and extracellular synthesis in accordance with the location where nanoparticles are synthesised. The intracellular method consists of transporting ions into the microbial cell to form nanoparticles in the presence of bio-catalysts. The extracellular biosynthesis of nanoparticles involves trapping the metal ions on the surface of the cells and reducing ions in the presence of enzymes. The bio-synthesized nanoparticles have been used in a variety of applications, including drug carriers for targeted delivery, cancer treatment, gene therapy and

DNA analysis, antibacterial agents, biosensors, enhancing reaction rates, separation science, and magnetic resonance imaging (MRI).

3. Characterization of Nanoparticles

It is very important to characterize the size, crystal structure, elemental composition, and physical properties of nanoparticles through various techniques. X-ray, spectroscopy and high-resolution microscopy techniques are normally used to measure the size and morphology of nanoparticles at sub-nanometer resolution [19]. XRD (X Ray diffraction) provides info pertaining to the crystalline structure, nature of the phase of particles, its lattice parameters and crystalline grain size [26].

Microscopy methods can create images of individual nanoparticles (NPs) to point out their shape, size and even the location. Because nanoparticles (NPs) have a size below the diffraction limit of visible light, regular light microscopic techniques are not effective. Electron microscopy (EMs) can go hand in hand with spectroscopic methods for performing elemental analysis of the Nanoparticles. Microscopy methods are highly destructive, further they can be very prone to undesirable artifacts/defects from sample preparation (drying or vacuum conditions), or from probe tip geometry (like in scanning probe microscopy). To add to the woes, microscopic techniques are based on single-particle measurements, which means that large numbers of individual particles must be enumerated together to estimate their bulk properties [27, 28]. An updated method where enhanced dark-field microscopy with hyperspectral imaging was used to show promise for imaging of nanoparticles (NPs) at complex matrices such as biological samples which has higher contrast and throughput Spectroscopy, which measures the particles' interaction with electromagnetic radiation (EMs) as a function of their wavelength, is useful for some classes of nanoparticles to characterize concentration (conc.), size and shape. The Semiconductor quantum dots are fluorescent and metal nanoparticles (NPs) exhibit surface plasmon absorbances, making both amenable to ultraviolet–visible (UV-VIS) spectroscopy. Infrared, nuclear magnetic resonance (NMR), along with X-ray spectroscopy are also used with nanoparticles. Light scattering methods using laser light, X-rays, or neutron scattering are used to determine particle size, with each method suitable for different size ranges and particle constituents [27,28].

Also, heterogeneous methods include electrophoresis for surface charge, the Brunauer–Emmett–Teller method for surface area, and X-ray diffraction for crystal structure as well as mass spectrometry (MS) for particle mass, and particle counters for particle number. Common chemical methods like Chromatography, centrifugation, and filtration techniques also can be used to distinguish nanoparticles in accordance to size or other physical properties before or during their enumeration or characterization [27,28].

3.1 Size and Shape Analysis

Size and shape characterization provide insight into the relationship between structure and properties of nanoparticles, which estimates their potential application in the field of modern biotechnology [29]. Therefore, investigating their morphological features is a critical strategy to be employed before finding any biological application. NPs are the most efficient and extensively employed drug delivery vehicle. Triggering the immune system is a major implication that needs higher concern in the design and synthesis of such vehicles [30]. Since immune responses rely on size dependent manners, NPs are designed accordingly with a size less than 5 nm as a vehicle which also helps in achieving an appropriate renal clearance [31]. Uptake into cells is efficiently achieved through maintaining a size range of 30-50 nm [32]. Larger NPs with a size of 100 nm and above are extensively employed in delivering multiple drugs to their target sites [33]. Internalization into cells via clathr-mediated endocytosis depends on the particles' size, including a size range of 30-200 nm diameter [34]. Enhanced blood circulation half-life and decreased accumulation in the liver were displayed by a gold NP with a size of 10 nm compared to 20 nm size [35].

A wide range of studies determined the influence on the efficiency of NPs by modulating their shape. The interaction of NPs with cell membranes is determined by their shape. A study focusing on the effect of NP geometry in determining mammalian cell uptake reported that nanorods were preferentially internalized than nanodiscs [36]. Zheng et al. reported that the nanorods displayed an enhanced cellular internalization and bioavailability when compared with nanospheres. Biodistribution study of Magnetic Mesoporous Silica Nanoparticles (M-MSNPs) in mice reported that the rod-like particles accumulated in tumor sites preferentially than spheres, which indicated an enhanced efficacy of drug delivery for cancer therapy. Hence, it is evident that designing NPs with suitable shape and size is crucial in finding the desired application in biology [37].

There are many sophisticated instruments for particle size and shape analysis. Atomic Force Microscopy (AFM) offers high-resolution imaging, measuring shapes, sizes, roughness, and nanomechanical characterization of drug delivery vehicles [38]. Scanning Electron Microscopy (SEM) could be employed for image acquisition by employing reflective and transmitted working modes [39]. Transmission Electron Microscopy provides 2D projected images of the particles at a required scale. Complications may arise during the measurement of the actual volume of a 3D object using a 2D projected image. Underestimation of such measurements is a major limitation in deliberating the reliability of the particle characterization. Such limitations could be deciphered through Electron Tomography, a sophisticated instrument employed for quantitatively measuring the shape and size of 3D objects [40].

3.2   Surface Charge and Zeta Potential

Zeta potential represents the surface charge of NPs, which is an important characteristic that affects their stability, aggregation and toxicity [41]. Depending on the surface chemistry of the particles, zeta potential can be positive, negative or neutral, and they can be controlled by modifying their surface chemistry with functional groups such as carboxyl, amino or hydroxyl groups [42].

Zeta potential could be employed as a tool in cancer therapy since it provides substantial knowledge and insight into the interaction of the particles with normal and cancer cells. This enables drugs to reach the specific site as the NPs recognize the cancer cells from normal human cells [43]. NPs are applied as sensors in the medical field for imaging and implant coatings. Surface charge is a crucial factor affecting such functions by influencing the biocompatibility and stability of the NPs [44]. Drug delivery to tumor sites can be enhanced by controlling their surface charge by activating the neutral NPs to cationic NPs in acidic tumor environments, which could be suitable for efficient uptake by the cells with negative membrane potential [45]. NPs with positive or negative charges could be a promising vehicle for drug delivery in dermal applications. A study shows the enhanced binding efficiency of charged particles in inflamed skin and reduced alkaline phosphatase activity [46]. Unexpected cytotoxicity may be caused by administering positive charged NPs through protein opsonization and non-specific cellular interaction. This could be prevented by employing less negative anionic particles with smaller sizes [47]. Therefore, zeta potential characterization should be incorporated into the process of developing suitable NPs. Due to prominent sensitivity, accuracy, and adaptability, electrophoretic light scattering (ELS), acoustic, and electroacoustic are the three methods for determining the zeta potential of suspended particles and ELS is the most effective strategy among them [48]. Since its development and commercialization in the last twenty years of the twentieth century, the electroacoustic method has consistently shown its capacity to assess the zeta potential and particle size of a wide variety of aqueous, concentrated, colloidal suspensions, and emulsions. The interparticle forces can be controlled to create stable suspensions and produce improved-quality products by precisely monitoring the isoelectric point [49].

3.3   Therapeutical characterization

Polyelectrolyte complex, Ionotropic gelation, Micro emulsion, Covalent cross-linking, Incorporation and incubation, Solvent evaporation, Coprecipitation, Complex-coacervation methods are different ways for chitosan nanoparticles to effectively deliver drug/genes to target sites [50]. The two methods utilized to create micro- and nanoparticle drug carriers are top-down and bottom-up methods. In the latter, a state of molecular dispersion type is used to prepare the particulate system, which is then permitted to associate with the ensuing production of solid particles. In contrast to top-down procedures, which start with large-scale materials and break them down into smaller particles, bottom-up techniques aim to arrange smaller components into assemblies of complicated structure. Traditional nanoparticle synthesis often utilizes bottom-up methods [51].

The most popular physical cross-linking technique is ionic cross-linking since it is easy to prepare, doesn't require an organic solvent or high temperatures, and doesn't involve any chemical interactions. Ionic gelation is the process in which an anionic cross-linker, such as sodium sulfate or tripolyphosphate (TPP) is employed [52]. In the chemical cross-linking method, an interaction is achieved between cross-linking agent and amino group present in chitosan [53]. In the solvent evaporation method the required drug is prepared as microspheres after evaporating previously emulsified volatile solvent [54].

4.   Applications of Nanoparticles in Biology

NPs have a wide range of individual, multimodal, and multifunctional applications both *in vitro* and *in vivo*, including biomolecular recognition, therapeutic delivery, biosensing, and bioimaging [55]. A number of companies like Advectus Life Sciences Inc., Alnis Biosciences, Inc., CapsulutionNanoScience AG, Eiffel Technologies, NanoPharm AG, NanoBio Corporation, PSiVida Ltd are involved in the development and commercialisation of nanomaterials in biological and medical applications [56]. Majority of these companies have exploited NPs in pharmaceuticals for drug delivery.

4.1 Imaging and Detection

Nanoparticle-based agents are employed in most common biomedical imaging modalities such as fluorescence imaging, MRI, CT, US, PET and SPECT [57]. Biological events monitoring and visualizing techniques should be highly sensitive and stable for desired response and results. Imaging techniques utilizing NPs doped with fluorescent dyes are gaining increased interest as they can selectively recognize the target with improved internalization for proper signalling and spatial response [58]. Enhancement of resolution can be achieved through surface, charge and structure modifications of the particles, combined with the utilization of super-resolution fluorescence microscopy [59]. The promising efficacy of NPs in fluorescent imaging instigated further studies in cancer diagnosis and treatment to ensure a precise result through multifunctional photo theranostics with combined modalities [60].

Radiotherapy for cancer is guided through MRI detection, which requires an acquired image for defined detection of the desired structure. Advancement in the development of gadolinium-based NPs resolves such issues by offering enhanced acquired images [61]. Computed Tomography (CT) guided devices are employed for the detection of various internal injuries and tumor development. Prompt diagnosis can be easily achieved through higher contrast images, which can be provided by the utilization of higher atomic number metal NPs that act as contrasting agents [62]. Single photon emission computed tomography (SPECT) can evaluate myocardial infusion and optimally track cells for preclinical evaluations. For such biomedical applications NPs are exploited as cell trackers for potential prognosis [63]. Also, various consequences are faced through the absence of efficient and fast diagnosis for cancers. Magnetic Nanoparticles (MNPs) overcome such challenges and open a possibility for early detection [64]. Satisfactory clinical outcomes are gained through introducing Deep brain stimulation (DBS), which has an impact on axons and dendrites. Challenges and limitations caused by such implantation procedures instigated for finding minimally invasive strategies with lower side effects. NPs are a prominent solution for many of the problems imparted by DBS. NPs with potential stimulation parameters are promising candidates for the management and treatment of many diseases and also have the ability to cross barriers to act on multiple targets [65].

Other molecular imaging techniques like confocal laser scanning microscopy (CLS), electroencephalography (EEG), fluorescence resonance energy transfer (FRET), magnetoencephalography (MEG) and transcranial magnetic stimulation (TMS) are also in practice [66]. Surface plasmon resonance (SPR) enhanced light scattering and absorption of gold nanoparticles allows molecular-specific imaging as well as detection of cancer and is also exploited for the selective laser photothermal therapy of cancer [67].

4.2 Drug Delivery

Metal nanoparticles are extensively employed for delivering various therapeutic agents to the desired sites. Properties of the nanoparticles, such as optical and surface properties, efficiently contribute to making them suitable carriers of drugs and other agents [68,69]. NPs can specifically target damaged or diseased tissues, which is an efficient characteristic for a successful drug delivery [70]. The possibility of customizing drug release and bioavailability profiles further enhanced their potential for selection as an incomparable candidate for utilization in drug delivery with minimal impact [71]. 50 NP-based drugs are approved by the FDA, which implies the tremendous capability of NPs in delivering drugs to specific targets [72].

Since the last two decades, lipids NPs are gaining much attention because of their biocompatibility and non-toxicity. Hybrid lipid-polymeric nanoparticles are a promising recently reported novel delivery system [73]. Microfluidics presents a viable method for producing NPs for medication delivery due to its capacity to precisely manipulate the characteristics of NPs. Making different NPs using microfluidics has a lot of potential, especially for those based on single-step synthesis methods. For instance, mixing-induced nanoprecipitation is used in microfluidic systems to create the majority of organic NPs [74]. Intense research is being done on nanoparticle-based drug carriers to get beyond the skin barrier and enhance even hydrophilic or macromolecular drug delivery into or across the skin effectively. Due to their distinctive qualities and adaptability, the use of gold nanoparticles (GNPs) as a novel type of drug carrier for cutaneous drug administration has gained attention during the past several years. Plasmonic properties and their biocompatibility add additional advantages as a drug delivery vehicle in skin [75]. Mesoporous silica NPs is another type of carrier for precisely locating body targets. Pharmaceutical agents are efficiently packed into the materials since they offer a porous structure, renal clearance, and optimum bio-distribution [76]. NPs have several applications nowadays. They have been investigated as imaging contrast agents, tumor gene delivery systems, and medication transporters. Enhancing cancer therapy has involved the use of nanomaterials made of organic, inorganic, lipid, or glycan compounds, as well as synthetic polymers [77].

4.3   Gene therapy

The treatment of hereditary illnesses with gene therapy has considerable potential with the help of the CRISPR system. Distribution of these CRISPR-derived technologies to particular organs, however, still poses a significant challenge. A very promising delivery mechanism has evolved in the form of lipid nanoparticles (LNPs) with various formulations for targeting different organs [78]. Retinoblastoma is an ocular cancer that is an important obstacle to health in different countries owing to its poor prognosis in its early stages. NPs such as gold, lipid, and polymeric based systems have displayed efficient gene delivery [79]. Inefficient delivery of nucleic acid based therapeutic systems instigated a higher need for efficient delivery systems for gene therapy. A recent finding reported that engineered NPs can optimally deliver nucleic acid into subcellular compartments [80].

4.4   Biosensors

Detection of pathogens is vital for preventing many life-threatening diseases. Gold NPs are incorporated into existing biosensing systems for enhanced detection of pathogens [81]. Increased recognition and detection could be possible by utilizing localised surface plasmon resonance (LSPR) biosensors which is based on silver and gold NPs [82]. Graphene oxide nanoparticles are developed for detecting the presence of oxalate in plasma and urine [83]. The luminescence resonance energy transfer (LRET) method is employed to detect trace elements in physiological environments with the development of upconversion-nanoparticle (UCNPs) based probes [84]. Li Nanodiagnostic technology mostly employs metal NPs in biomedical applications because of their suitable optoelectronic properties which contribute to their efficient signal transduction [85]. For fast response and enhanced target specificity aptamer conjugated gold nanoparticles are developed with numerous applications, which include cancer and other biological component detections [86].

4.5   Tissue engineering

The prospect of inducing new functional bone regeneration is provided by tissue engineering techniques, with the biomimetic scaffold acting as a link to establish a microenvironment that permits the emergence of a regenerative niche at the site of injury. The use of magnetic nanoparticles in bone tissue engineering has the potential to enhance the osteoinductive, osteoconductive, and angiogenic properties of scaffolds by taking advantage of the magnetism that magnetic nanoparticles naturally possess in cellular microenvironments [87]. Nanoporous scaffolds and nanofiber membranes are just two examples of the biocompatible nanomaterials that may now be used in tissue engineering because of the rapidly developing nanotechnological and manufacturing techniques [88]. Mesoporous silica NPs incorporated as scaffold in bone tissue engineering resulted in favourable results such as improved mechanical strength of the scaffold, desired swelling behaviour and decreased hydrolytic degradation without any cytotoxic effects [89].

4.6   Cancer treatment

Good biocompatibility and controlled biodistribution patterns of functionalized gold nanoparticles make them excellent candidates for the foundation of novel therapeutics. The spectrum of their possible medicinal uses, with a focus on cancer treatment, is greatly expanded by the ability to change the surface of particles with various targeting and functional chemicals [90]. For the creation and advancement of novel cancer therapies, a variety of nanomaterials based on organic as well as on synthetic polymers have been used [91]. Nanoparticles have the benefit of passively targeting cancer by amassing and becoming trapped in tumors. The enhanced permeation and retention effect is a phenomenon that has been used to explain why macromolecules and nanoparticle ratios are higher in tumors than in healthy tissues [92]. Stimuli-responsive NPs could be considered a novel system for drug delivery since they release the drug only upon homing to the cancer environment in response to specific signals

4.7   Antimicrobial agents

Many NPs exhibit antimicrobial properties, which is an essential quality for a modern medical device. Synergistic effects can be observed when the currently employed antibiotics are combined with the metal nanoparticles (MNPs) [93]. Their characteristic features such as selective recognition, non-cytotoxicity and high surface area render these particles more effective for exhibiting biological action [94]. Natural products are the commonly employed agents for antimicrobial effects, and they face many challenges. Combining NPs with such products results in promising effects against gram positive and gram negative bacteria [95]. The most promising approach to improve food quality and safety is the use of essential oils (EO) that have been loaded with nanoparticles. The antibacterial capabilities of EO for pathogen control in natural and processed foods for human health and animal production are described in several works [96]. Extensive need for biomaterials for medical needs, such as bandages and wound dressing, instigated the development of such products. Application of silk fibroin (SF), a nanofibre derived from silkworm with encased silver NPs is a developed strategy for such a problem. NPs confer antimicrobial effect on SF [10].

Nanomaterials in Biological Systems   Materials Research Forum LLC
Materials Research Foundations 185 (2026) 35-   https://doi.org/10.21741/9781644903858-2

5.   Case Studies of Nanoparticles Applications in Biology

Magnetic nanoparticles are the most attractive agent employed for cancer treatment because of their intrinsic property. Iron nanoparticles (FeNPs) possess superior magnetic and biological properties among all other studied MNPs, and hence, they are extensively employed for treatment and diagnosis. A recent study conducted on evaluating the cytotoxicity of multilayered graphene encapsulated Fe NPs resulted in a significant reduction in cell growth and revealed the potential of such systems in biomedicine [97]. Due to their outstanding MRI performance, long blood circulation time after proper surface modification, renal clearance capacity, and remarkable biosafety profile, ultrasmall superparamagnetic iron oxide (USPIO) nanoparticles with a core diameter of less than 5.0 nm are anticipated to become the next generation of contrast agents in MRI imaging [98].

A recent study described the microwave assisted synthesis of highly fluorescent nitrogen-doped graphene quantum dots (N-GQDs) from waste precursors and their functional characterization. The application of the prepared product in breast cancer cell line revealed a promising result in imaging since nucleic acids were observed in blue fluorescent colour [99].

A study focused on the efficacy of ionizable lipid NPs in delivering mRNA to pancreatic β cells. The result demonstrates that the intraperitoneal delivery of cationic helper lipid-containing lipid nanoparticles results in robust and targeted protein expression in the pancreatic cells. Therefore, this approach will make it possible to treat stubborn pancreatic disorders like diabetes and cancer with gene therapies [100].

In a study that focused on determining the potential wound-healing effect of cellulose NPs characterized the physicochemical properties of Electrospun fibers made of polyurethane and cellulose acetate that contain rosemary essential oil and silver nanoparticles that have been absorbed. The enhanced hydrophilicity of the fiber was evident in the cell viability and attachment assay, which led to better cell adhesion to the micro-nanofibers in a manner similar to the body's natural extracellular matrix [101].

Antimicrobial assessment of silver NPs was carried out against clinical isolates in a recent study. The study shows a Synergistic effect of antimicrobial action of AgNPs when combined with conventional antibiotics against gram-negative strains without displaying any cytotoxicity towards hepatocytes. Therefore, it is evident that they have the potential to act as antimicrobial agents without harmful effects [102].

A study was conducted to find a competent target precise NPs that carry the drug to ovarian cancer cells. Diverse polymeric nanoparticles were compared for anticancer activity and biocompatibility. Among them, chitosan and poly sarcosine nanoparticles (CS-PSar-NPs) displayed a prominent response [103].

NPs, especially the metallic ones, are used to produce biofuels such as biodiesel, biohydrogen, biogas, and bioethanol which can improve production performance and efficiency [104]. Biomolecules and drug-functionalized AuNPs are effectively used to treat various cancers and other diseases [105].

6. Challenges and Future Directions of Nanoparticle Applications in Biology

Humans are highly exposed to Nanoparticles nowadays since childhood, and they may create hazardous effects through increased exposure as it is rapidly developing [106]. Because of their smaller size they may reach sensitive organs easily and may cause harmful effects [107]. Chen et al. reported that aluminium based nanoparticles disrupt various physiological pathways and also effects cell viability. Usually, gold NPs are safe and non-toxic for use as a delivery vehicle, some reports suggest that they possess limited toxicity in the dose and side chain dependent manner. Cytotoxic assessment of copper NPs in liver and kidney reported that they possess toxic effects. Many other kinds of NPs subjected to toxicological analysis revealed that many of them possess toxicity in a dose and chemistry dependent manner [108]. Regulatory bodies have developed strategies to include nanomaterials into the existing legislation of chemicals, biocides, and food additives. The evaluation of toxicological effects had to be expanded from target organ effects (e.g., lungs, skin) to potential systemic effects. Due to the tiny size and an increased dissolution of nanoparticles, the toxicokinetics became a predominant additional endpoint. The main regulatory bodies such as the Organisation for Economic Co-operation and Development (OECD), European Chemicals Agency (ECHA), European Food Safety Authority (EFSA), European Medicines Agency (EMA), US Food and Drug Administration (FDA), and US Environmental Protection Agency (EPA) have issued technical guidelines, decisions, and laws to regulate nanomaterials In order to break down biological barriers and improve medication targeting to damaged tissue as well as reduce drug accumulation in non-specific organs, physicochemical qualities of MNPs should also be given higher priority [109]. Metal nanoparticle aggregation problems need solutions that encourage thermal breakdown. In order to facilitate thermal decomposition and make it simpler to manufacture monodisperse nanoparticles, the etching synthesis method of metal nanoparticles will be promoted [110].

Conclusion

Nanotechnology, being a rapidly developing field, they provides an elaborate opportunity in targeted drug delivery, gene therapy, tissue engineering, biosensor development, and antimicrobial treatments. In this review, the characterization and application of nanoparticles were evaluated in detail. Much attention was received in characterizing NPs as it determines the therapeutic potential and efficiency of the particles, and considering these aspects, suitable methods are available for synthesizing NPs. NPs could clear major obstructions in medical science through their multifunctional and non-toxic properties. Poor prognosis is majorly responsible for the socio-economic burden caused by various ailments. Capabilities of NPs particles in biocompatibility, cell uptake, and attachment with biological components enable them to act as a major candidate in imaging technology. Therefore, it is evident that the development of nanoparticles provides promising progress in disease diagnosis and treatment. From the analysis of the different case studies, it is concluded that NPs could be a sustainable solution for the majority of the implications in treatment-associated problems.

References

[1] V. Chandrakala, V. Aruna, G. Angajala, Review on metal nanoparticles as nanocarriers: current challenges and perspectives in drug delivery systems, Emergent Mater. 5 (2022) 1593–1615. https://doi.org/10.1007/s42247-021-00335-x.

[2]   H. Xiang, J. Meng, W. Shao, D. Zeng, J. Ji, P. Wang, X. Zhou, P. Qi, L. Liu, S. Yang, Plant protein-based self-assembling core–shell nanocarrier for effectively controlling plant viruses: Evidence for nanoparticle delivery behavior, plant growth promotion, and plant resistance induction, Chem. Eng. J. 464 (2023). https://doi.org/10.1016/j.cej.2023.142432.

[3]   E. Roduner, Size matters: Why nanomaterials are different, Chem. Soc. Rev. 35 (2006). https://doi.org/10.1039/b502142c.

[4]   E.C. Dreaden, A.M. Alkilany, X. Huang, C.J. Murphy, M.A. El-Sayed, The golden age: Gold nanoparticles for biomedicine, Chem. Soc. Rev. 41 (2012). https://doi.org/10.1039/c1cs15237h.

[5]   I. Khan, K. Saeed, I. Khan, Nanoparticles: Properties, applications and toxicities, Arab. J. Chem. 12 (2019). https://doi.org/10.1016/j.arabjc.2017.05.011.

[6]   W.K. Shin, J. Cho, A.G. Kannan, Y.S. Lee, D.W. Kim, Cross-linked Composite Gel Polymer Electrolyte using Mesoporous Methacrylate-Functionalized SiO2 Nanoparticles for Lithium-Ion Polymer Batteries, Sci. Rep. 6 (2016). https://doi.org/10.1038/srep26332.

[7]   A.M. Ealias, M.P. Saravanakumar, A review on the classification, characterisation, synthesis of nanoparticles and their application, in: IOP Conf. Ser. Mater. Sci. Eng., 2017. https://doi.org/10.1088/1757-899X/263/3/032019.

[8]   G. Romero, S.E. Moya, Synthesis of organic nanoparticles, in: Front. Nanosci., 2012. https://doi.org/10.1016/B978-0-12-415769-9.00004-2.

[9]   G. Karunakaran, K.G. Sudha, S. Ali, E.B. Cho, Biosynthesis of Nanoparticles from Various Biological Sources and Its Biomedical Applications, Molecules 28 (2023). https://doi.org/10.3390/molecules28114527.

[10] R.S. Khan, A.H. Rather, T.U. Wani, S. ullah Rather, A. Abdal-hay, F.A. Sheikh, A comparative review on silk fibroin nanofibers encasing the silver nanoparticles as antimicrobial agents for wound healing applications, Mater. Today Commun. 32 (2022). https://doi.org/10.1016/j.mtcomm.2022.103914.

[11] M. Xu, W. Yim, J. Zhou, J. Zhou, Z. Jin, C. Moore, R. Borum, A. Jorns, J. V. Jokerst, The Application of Organic Nanomaterials for Bioimaging, Drug Delivery, and Therapy: Spanning Various Domains, IEEE Nanotechnol. Mag. 15 (2021). https://doi.org/10.1109/MNANO.2021.3081758.

[12] D. Holmannova, P. Borsky, T. Svadlakova, L. Borska, Z. Fiala, Carbon Nanoparticles and Their Biomedical Applications, Appl. Sci. 12 (2022). https://doi.org/10.3390/app12157865.

[13] D. Maiti, X. Tong, X. Mou, K. Yang, Carbon-Based Nanomaterials for Biomedical Applications: A Recent Study, Front. Pharmacol. 9 (2019). https://doi.org/10.3389/fphar.2018.01401.

[14] C. Altavilla, E. Ciliberto, Inorganic Nanoparticles: Synthesis, Applications, and Perspectives, 2017. https://doi.org/10.1201/b10333.

[15] S. Kaushik, Polymeric and Ceramic Nanoparticles: Possible Role in Biomedical Applications, in: Handb. Polym. Eramic Nanotechnol. Vol. 1,2, 2021: pp. 1293–1308. https://doi.org/10.1007/978-3-030-40513-7_39.

[16]   M.H. Sarfraz, M. Zubair, B. Aslam, A. Ashraf, M.H. Siddique, S. Hayat, J.N. Cruz, S.
        Muzammil, M. Khurshid, M.F. Sarfraz, A. Hashem, T.M. Dawoud, G.D. Avila-Quezada,
        E.F. Abd_Allah, Comparative analysis of phyto-fabricated chitosan, copper oxide, and
        chitosan-based CuO nanoparticles: antibacterial potential against Acinetobacter baumannii
        isolates and anticancer activity against HepG2 cell lines, Front. Microbiol. 14 (2023)
        1188743. https://doi.org/10.3389/FMICB.2023.1188743/BIBTEX.

[17]   J.N. Cruz, S. Muzammil, A. Ashraf, M.U. Ijaz, M.H. Siddique, R. Abbas, M. Sadia, Saba,
        S. Hayat, R.R. Lima, A review on mycogenic metallic nanoparticles and their potential
        role as antioxidant, antibiofilm and quorum quenching agents, Heliyon 10 (2024) e29500.
        https://doi.org/10.1016/J.HELIYON.2024.E29500.

[18]   I. Ijaz, E. Gilani, A. Nazir, A. Bukhari, Detail review on chemical, physical and green
        synthesis, classification, characterizations and applications of nanoparticles, Green Chem.
        Lett. Rev. 13 (2020). https://doi.org/10.1080/17518253.2020.1802517.

[19]   P. Szczyglewska, A. Feliczak-Guzik, I. Nowak, Nanotechnology–General Aspects: A
        Chemical Reduction Approach to the Synthesis of Nanoparticles, Molecules 28 (2023).
        https://doi.org/10.3390/molecules28134932.

[20]   S. Iravani, Methods for Preparation of Metal Nanoparticles, in: Met. Nanoparticles, 2018.
        https://doi.org/10.1002/9783527807093.ch2.

[21]   F.S. Alves, J. de A. Rodrigues Do Rego, M.L. Da Costa, L.F. Lobato Da Silva, R.A. Da
        Costa, J.N. Cruz, D.D.S.B. Brasil, Spectroscopic methods and in silico analyses using
        density functional theory to characterize and identify piperine alkaloid crystals isolated
        from pepper (Piper Nigrum L.), J. Biomol. Struct. Dyn. 38 (2020) 2792–2799.
        https://doi.org/10.1080/07391102.2019.1639547.

[22]   A. Ghadi, S. Mahjoub, F. Tabandeh, F. Talebnia, Synthesis and optimization of chitosan
        nanoparticles: Potential applications in nanomedicine and biomedical engineering, Casp.
        J. Intern. Med. 5 (2014).

[23]   S. Muzammil, J. Neves Cruz, R. Mumtaz, I. Rasul, S. Hayat, M.A. Khan, A.M. Khan,
        M.U. Ijaz, R.R. Lima, M. Zubair, Effects of Drying Temperature and Solvents on In Vitro
        Diabetic Wound Healing Potential of Moringa oleifera Leaf Extracts, Molecules 28 (2023)
        710. https://doi.org/10.3390/MOLECULES28020710/S1.

[24]   I.N. de F. Ramos, M.F. da Silva, J.M.S. Lopes, J.N. Cruz, F.S. Alves, J. de A.R. do Rego,
        M.L. da Costa, P.P. de Assumpção, D. do S. Barros Brasil, A.S. Khayat, Extraction,
        Characterization, and Evaluation of the Cytotoxic Activity of Piperine in Its Isolated form
        and in Combination with Chemotherapeutics against Gastric Cancer, Mol. 2023, Vol. 28,
        Page 5587 28 (2023) 5587. https://doi.org/10.3390/MOLECULES28145587.

[25]   S. Mandal, R. Krishnan, Fungi: The budding source for biomaterials, Microb. Biosyst. 6
        (2021). https://doi.org/10.21608/mb.2021.92305.1038.

[26]   S. Mourdikoudis, R.M. Pallares, N.T.K. Thanh, Characterization techniques for
        nanoparticles: comparison and complementarity upon studying nanoparticle properties,
        Nanoscale 10 (2018) 12871–12934. https://doi.org/10.1039/C8NR02278J.

[27]   M. Hassellöv, J.W. Readman, J.F. Ranville, K. Tiede, Nanoparticle analysis and
        characterization methodologies in environmental risk assessment of engineered

nanoparticles, Ecotoxicology 17 (2008). https://doi.org/10.1007/s10646-008-0225-x.

[28] K. Tiede, A.B.A. Boxall, S.P. Tear, J. Lewis, H. David, M. Hassellöv, Detection and characterization of engineered nanoparticles in food and the environment, Food Addit. Contam. 25 (2008) 795–821.

[29] M. Kaliva, M. Vamvakaki, Nanomaterials characterization, in: Polym. Sci. Nanotechnol. Fundam. Appl., 2020. https://doi.org/10.1016/B978-0-12-816806-6.00017-0.

[30] Y. Liu, J. Hardie, X. Zhang, V.M. Rotello, Effects of engineered nanoparticles on the innate immune system, Semin. Immunol. 34 (2017). https://doi.org/10.1016/j.smim.2017.09.011.

[31] H. Soo Choi, W. Liu, P. Misra, E. Tanaka, J.P. Zimmer, B. Itty Ipe, M.G. Bawendi, J. V. Frangioni, Renal clearance of quantum dots, Nat. Biotechnol. 25 (2007). https://doi.org/10.1038/nbt1340.

[32] F. Lu, S.H. Wu, Y. Hung, C.Y. Mou, Size effect on cell uptake in well-suspended, uniform mesoporous silica nanoparticles, Small 5 (2009). https://doi.org/10.1002/smll.200900005.

[33] Y.L. Su, J.H. Fang, C.Y. Liao, C.T. Lin, Y.T. Li, S.H. Hu, Targeted mesoporous iron oxide nanoparticles-encapsulated perfluorohexane and a hydrophobic drug for deep tumor penetration and therapy, Theranostics 5 (2015). https://doi.org/10.7150/thno.12843.

[34] J. Rejman, V. Oberle, I.S. Zuhorn, D. Hoekstra, Size-dependent internalization of particles via the pathways of clathrin-and caveolae-mediated endocytosis, Biochem. J. 377 (2004). https://doi.org/10.1042/BJ20031253.

[35] L. Talamini, M.B. Violatto, Q. Cai, M.P. Monopoli, K. Kantner, Ž. Krpetić, A. Perez-Potti, J. Cookman, D. Garry, C.P. Silveira, L. Boselli, B. Pelaz, T. Serchi, S. Cambier, A.C. Gutleb, N. Feliu, Y. Yan, M. Salmona, W.J. Parak, K.A. Dawson, P. Bigini, Influence of Size and Shape on the Anatomical Distribution of Endotoxin-Free Gold Nanoparticles, ACS Nano 11 (2017). https://doi.org/10.1021/acsnano.7b00497.

[36] R. Agarwal, V. Singh, P. Jurney, L. Shi, S. V. Sreenivasan, K. Roy, Mammalian cells preferentially internalize hydrogel nanodiscs over nanorods and use shape-specific uptake mechanisms, Proc. Natl. Acad. Sci. U. S. A. 110 (2013). https://doi.org/10.1073/pnas.1305000110.

[37] N. Zheng, J. Li, C. Xu, L. Xu, S. Li, L. Xu, Mesoporous silica nanorods for improved oral drug absorption, Artif. Cells, Nanomedicine Biotechnol. 46 (2018). https://doi.org/10.1080/21691401.2017.1362414.

[38] A. Sarkar, Biosensing, Characterization of Biosensors, and Improved Drug Delivery Approaches Using Atomic Force Microscopy: A Review, Front. Nanotechnol. 3 (2022). https://doi.org/10.3389/fnano.2021.798928.

[39] A.E. Vladár, V.D. Hodoroaba, Characterization of nanoparticles by scanning electron microscopy, in: Charact. Nanoparticles Meas. Process. Nanoparticles, 2019. https://doi.org/10.1016/B978-0-12-814182-3.00002-X.

[40] M. Hayashida, F. Paraguay-Delgado, C. Ornelas, A. Herzing, A.M. Blackburn, B. Haydon, T. Yaguchi, A. Wakui, K. Igarashi, Y. Suzuki, S. Motoki, Y. Aoyama, Y. Konyuba, M. Malac, Nanoparticle size and 3D shape measurement by electron

tomography: An Inter-Laboratory Comparison, Micron 140 (2021). https://doi.org/10.1016/j.micron.2020.102956.

[41] J. Jiang, G. Oberdörster, P. Biswas, Characterization of size, surface charge, and agglomeration state of nanoparticle dispersions for toxicological studies, J. Nanoparticle Res. 11 (2009). https://doi.org/10.1007/s11051-008-9446-4.

[42] L. Abarca-Cabrera, P. Fraga-García, S. Berensmeier, Bio-nano interactions: binding proteins, polysaccharides, lipids and nucleic acids onto magnetic nanoparticles, Biomater. Res. 25 (2021). https://doi.org/10.1186/s40824-021-00212-y.

[43] Y. Zhang, M. Yang, N.G. Portney, D. Cui, G. Budak, E. Ozbay, M. Ozkan, C.S. Ozkan, Zeta potential: A surface electrical characteristic to probe the interaction of nanoparticles with normal and cancer human breast epithelial cells, Biomed. Microdevices 10 (2008). https://doi.org/10.1007/s10544-007-9139-2.

[44] E. Fröhlich, The role of surface charge in cellular uptake and cytotoxicity of medical nanoparticles, Int. J. Nanomedicine 7 (2012). https://doi.org/10.2147/IJN.S36111.

[45] Y.Y. Yuan, C.Q. Mao, X.J. Du, J.Z. Du, F. Wang, J. Wang, Surface charge switchable nanoparticles based on zwitterionic polymer for enhanced drug delivery to tumor, Adv. Mater. 24 (2012). https://doi.org/10.1002/adma.201202296.

[46] M.M.A. Abdel-Mottaleb, B. Moulari, A. Beduneau, Y. Pellequer, A. Lamprecht, Surface-charge-dependent nanoparticles accumulation in inflamed skin, J. Pharm. Sci. 101 (2012). https://doi.org/10.1002/jps.23282.

[47] S.M.A. Sadat, S.T. Jahan, A. Haddadi, S.M.A. Sadat, S.T. Jahan, A. Haddadi, Effects of Size and Surface Charge of Polymeric Nanoparticles on in Vitro and in Vivo Applications, J. Biomater. Nanobiotechnol. 7 (2016) 91–108. https://doi.org/10.4236/JBNB.2016.72011.

[48] R. Xu, Progress in nanoparticles characterization: Sizing and zeta potential measurement, Particuology 6 (2008). https://doi.org/10.1016/j.partic.2007.12.002.

[49] S. Mohammadi-Jam, K.E. Waters, R.W. Greenwood, A review of zeta potential measurements using electroacoustics, Adv. Colloid Interface Sci. 309 (2022). https://doi.org/10.1016/j.cis.2022.102778.

[50] K.C. Hembram, S. Prabha, R. Chandra, B. Ahmed, S. Nimesh, Advances in preparation and characterization of chitosan nanoparticles for therapeutics, Artif. Cells, Nanomedicine Biotechnol. 44 (2016). https://doi.org/10.3109/21691401.2014.948548.

[51] N. Abid, A.M. Khan, S. Shujait, K. Chaudhary, M. Ikram, M. Imran, J. Haider, M. Khan, Q. Khan, M. Maqbool, Synthesis of nanomaterials using various top-down and bottom-up approaches, influencing factors, advantages, and disadvantages: A review, Adv. Colloid Interface Sci. 300 (2022). https://doi.org/10.1016/j.cis.2021.102597.

[52] S. Mao, U. Bakowsky, A. Jintapattanakit, T. Kissel, Self-assembled polyelectrolyte nanocomplexes between chitosan derivatives and insulin, J. Pharm. Sci. 95 (2006). https://doi.org/10.1002/jps.20520.

[53] F. Bugamelli, M.A. Raggi, I. Orienti, V. Zecchi, Controlled insulin release from chitosan microparticles, Arch. Pharm. (Weinheim). 331 (1998). https://doi.org/10.1002/(SICI)1521-4184(199804)331:4<133::AID-ARDP133>3.0.CO;2-H.

[54] N. Garud, A. Garud, Preparation and in-vitro evaluation of metformin microspheres using non-aqueous solvent evaporation technique, Trop. J. Pharm. Res. 11 (2012). https://doi.org/10.4314/tjpr.v11i4.8.

[55] M. De, P.S. Ghosh, V.M. Rotello, Applications of nanoparticles in biology, Adv. Mater. 20 (2008). https://doi.org/10.1002/adma.200703183.

[56] O. V. Salata, Applications of nanoparticles in biology and medicine, J. Nanobiotechnology 2 (2004). https://doi.org/10.1186/1477-3155-2-3.

[57] X. Han, K. Xu, O. Taratula, K. Farsad, Applications of nanoparticles in biomedical imaging, Nanoscale 11 (2019). https://doi.org/10.1039/c8nr07769j.

[58] K. Wang, X. He, X. Yang, H. Shi, Functionalized silica nanoparticles: A platform for fluorescence imaging at the cell and small animal levels, Acc. Chem. Res. 46 (2013). https://doi.org/10.1021/ar3001525.

[59] R. Xu, H. Cao, D. Lin, B. Yu, J. Qu, Lanthanide-doped upconversion nanoparticles for biological super-resolution fluorescence imaging, Cell Reports Phys. Sci. 3 (2022). https://doi.org/10.1016/j.xcrp.2022.100922.

[60] X. Yin, Y. Cheng, Y. Feng, W.R. Stiles, S.H. Park, H. Kang, H.S. Choi, Phototheranostics for multifunctional treatment of cancer with fluorescence imaging, Adv. Drug Deliv. Rev. 189 (2022). https://doi.org/10.1016/j.addr.2022.114483.

[61] L. Smith, Z. Kuncic, H.L. Byrne, D. Waddington, Nanoparticles for MRI-guided radiation therapy: a review, Cancer Nanotechnol. 13 (2022). https://doi.org/10.1186/s12645-022-00145-8.

[62] D. Gupta, I. Roy, S. Gandhi, Metallic nanoparticles for CT-guided imaging of tumors and their therapeutic applications, OpenNano 12 (2023). https://doi.org/10.1016/j.onano.2023.100146.

[63] Ł. Tekieli, W. Szot, E. Kwiecień, A. Mazurek, E. Borkowska, Ł. Czyż, M. Dąbrowski, A. Kozynacka, M. Skubera, P. Podolec, M. Majka, M. Kostkiewicz, P. Musiałek, Single-photon emission computed tomography as a fundamental tool in evaluation of myocardial reparation and regeneration therapies, Postep. w Kardiol. Interwencyjnej 18 (2022). https://doi.org/10.5114/AIC.2023.124403.

[64] A. Rastogi, K. Yadav, A. Mishra, M.S. Singh, S. Chaudhary, R. Manohar, A.S. Parmar, Early diagnosis of lung cancer using magnetic nanoparticles-integrated systems, Nanotechnol. Rev. 11 (2022). https://doi.org/10.1515/ntrev-2022-0032.

[65] M. Kujawska, A. Kaushik, Exploring magneto-electric nanoparticles (MENPs) a platform for implanted deep brain stimulation, Neural Regen. Res. 18 (2023). https://doi.org/10.4103/1673-5374.340411.

[66] P. Padmanabhan, A. Kumar, S. Kumar, R.K. Chaudhary, B. Gulyás, Nanoparticles in practice for molecular-imaging applications: An overview, Acta Biomater. 41 (2016). https://doi.org/10.1016/j.actbio.2016.06.003.

[67] P.K. Jain, I.H. ElSayed, M.A. El-Sayed, Au nanoparticles target cancer, Nano Today 2 (2007). https://doi.org/10.1016/S1748-0132(07)70016-6.

[68] E.R. Cooper, Nanoparticles: A personal experience for formulating poorly water soluble

drugs, J. Control. Release 141 (2010). https://doi.org/10.1016/j.jconrel.2009.10.006.

[69]   K. Thanki, R.P. Gangwal, A.T. Sangamwar, S. Jain, Oral delivery of anticancer drugs: Challenges and opportunities, J. Control. Release 170 (2013). https://doi.org/10.1016/j.jconrel.2013.04.020.

[70]   B. Felice, M.P. Prabhakaran, A.P. Rodríguez, S. Ramakrishna, Drug delivery vehicles on a nano-engineering perspective, Mater. Sci. Eng. C 41 (2014). https://doi.org/10.1016/j.msec.2014.04.049.

[71]   K.B. Ramadi, Y.A. Mohamed, A. Al-Sbiei, S. Almarzooqi, G. Bashir, A. Al Dhanhani, D. Sarawathiamma, S. Qadri, J. Yasin, A. Nemmar, M.J. Fernandez-Cabezudo, Y. Haik, B.K. Al-Ramadi, Acute systemic exposure to silver-based nanoparticles induces hepatotoxicity and NLRP3-dependent inflammation, Nanotoxicology 10 (2016). https://doi.org/10.3109/17435390.2016.1163743.

[72]   C.M. Dawidczyk, C. Kim, J.H. Park, L.M. Russell, K.H. Lee, M.G. Pomper, P.C. Searson, State-of-the-art in design rules for drug delivery platforms: Lessons learned from FDA-approved nanomedicines, J. Control. Release 187 (2014). https://doi.org/10.1016/j.jconrel.2014.05.036.

[73]   L. Xu, X. Wang, Y. Liu, G. Yang, R.J. Falconer, C.-X. Zhao, Lipid nanoparticles for drug delivery, Adv. NanoBiomed Res. 2 (2022) 2100109. https://doi.org/10.1002/anbr.202100109.

[74]   Y. Liu, G. Yang, Y. Hui, S. Ranaweera, C.X. Zhao, Microfluidic Nanoparticles for Drug Delivery, Small 18 (2022). https://doi.org/10.1002/smll.202106580.

[75]   Y. Chen, X. Feng, Gold nanoparticles for skin drug delivery, Int. J. Pharm. 625 (2022). https://doi.org/10.1016/j.ijpharm.2022.122122.

[76]   M. Vallet-Regí, F. Schüth, D. Lozano, M. Colilla, M. Manzano, Engineering mesoporous silica nanoparticles for drug delivery: where are we after two decades?, Chem. Soc. Rev. 51 (2022). https://doi.org/10.1039/d1cs00659b.

[77]   H. Sabit, M. Abdel-Hakeem, T. Shoala, S. Abdel-Ghany, M.M. Abdel-Latif, J. Almulhim, M. Mansy, Nanocarriers: A Reliable Tool for the Delivery of Anticancer Drugs, Pharmaceutics 14 (2022). https://doi.org/10.3390/pharmaceutics14081566.

[78]   K. Godbout, J.P. Tremblay, Delivery of RNAs to Specific Organs by Lipid Nanoparticles for Gene Therapy, Pharmaceutics 14 (2022). https://doi.org/10.3390/pharmaceutics14102129.

[79]   M. Mandal, I. Banerjee, M. Mandal, Nanoparticle-mediated gene therapy as a novel strategy for the treatment of retinoblastoma, Colloids Surfaces B Biointerfaces 220 (2022). https://doi.org/10.1016/j.colsurfb.2022.112899.

[80]   L.M. Mollé, C.H. Smyth, D. Yuen, A.P.R. Johnston, Nanoparticles for vaccine and gene therapy: Overcoming the barriers to nucleic acid delivery, Wiley Interdiscip. Rev. Nanomedicine Nanobiotechnology 14 (2022). https://doi.org/10.1002/wnan.1809.

[81]   M. Hegde, P. Pai, M.G. Shetty, K.S. Babitha, Gold nanoparticle based biosensors for rapid pathogen detection: A review, Environ. Nanotechnology, Monit. Manag. 18 (2022). https://doi.org/10.1016/j.enmm.2022.100756.

[82] Y. Ziai, C. Rinoldi, P. Nakielski, L. De Sio, F. Pierini, Smart plasmonic hydrogels based on gold and silver nanoparticles for biosensing application, Curr. Opin. Biomed. Eng. 24 (2022). https://doi.org/10.1016/j.cobme.2022.100413.

[83] R. Devi, S. Relhan, C.S. Pundir, Construction of a chitosan/polyaniline/graphene oxide nanoparticles/ polypyrrole/Au electrode for amperometric determination of urinary/plasma oxalate, Sensors Actuators, B Chem. 186 (2013). https://doi.org/10.1016/j.snb.2013.05.078.

[84] Y. Li, C. Chen, F. Liu, J. Liu, Engineered lanthanide-doped upconversion nanoparticles for biosensing and bioimaging application, Microchim. Acta 189 (2022). https://doi.org/10.1007/s00604-022-05180-1.

[85] S. Malathi, I. Pakrudheen, S.N. Kalkura, T.J. Webster, S. Balasubramanian, Disposable biosensors based on metal nanoparticles, Sensors Int. 3 (2022). https://doi.org/10.1016/j.sintl.2022.100169.

[86] S. Nooranian, A. Mohammadinejad, T. Mohajeri, G. Aleyaghoob, R. Kazemi Oskuee, Biosensors based on aptamer-conjugated gold nanoparticles: A review, Biotechnol. Appl. Biochem. 69 (2022). https://doi.org/10.1002/bab.2224.

[87] A. Dasari, J. Xue, S. Deb, Magnetic Nanoparticles in Bone Tissue Engineering, Nanomaterials 12 (2022). https://doi.org/10.3390/nano12050757.

[88] V. Harish, D. Tewari, M. Gaur, A.B. Yadav, S. Swaroop, M. Bechelany, A. Barhoum, Review on Nanoparticles and Nanostructured Materials: Bioimaging, Biosensing, Drug Delivery, Tissue Engineering, Antimicrobial, and Agro-Food Applications, Nanomaterials 12 (2022). https://doi.org/10.3390/nano12030457.

[89] S. Yousefiasl, H. Manoochehri, P. Makvandi, S. Afshar, E. Salahinejad, P. Khosraviyan, M. Saidijam, S. Soleimani Asl, E. Sharifi, Chitosan/alginate bionanocomposites adorned with mesoporous silica nanoparticles for bone tissue engineering, J. Nanostructure Chem. 13 (2023). https://doi.org/10.1007/s40097-022-00507-z.

[90] K. Sztandera, M. Gorzkiewicz, B. Klajnert-Maculewicz, Gold Nanoparticles in Cancer Treatment, Mol. Pharm. 16 (2019). https://doi.org/10.1021/acs.molpharmaceut.8b00810.

[91] A. Aghebati-Maleki, S. Dolati, M. Ahmadi, A. Baghbanzadeh, M. Asadi, A. Fotouhi, M. Yousefi, L. Aghebati-Maleki, Nanoparticles and cancer therapy: Perspectives for application of nanoparticles in the treatment of cancers, J. Cell. Physiol. 235 (2020). https://doi.org/10.1002/jcp.29126.

[92] M. Wang, M. Thanou, Targeting nanoparticles to cancer, Pharmacol. Res. 62 (2010). https://doi.org/10.1016/j.phrs.2010.03.005.

[93] A.I. Ribeiro, A.M. Dias, A. Zille, Synergistic Effects between Metal Nanoparticles and Commercial Antimicrobial Agents: A Review, ACS Appl. Nano Mater. 5 (2022). https://doi.org/10.1021/acsanm.1c03891.

[94] D.A. Mercan, A.G. Niculescu, A.M. Grumezescu, Nanoparticles for Antimicrobial Agents Delivery—An Up-to-Date Review, Int. J. Mol. Sci. 23 (2022). https://doi.org/10.3390/ijms232213862.

[95] D. Susanti, M.S. Haris, M. Taher, J. Khotib, Natural Products-Based Metallic Nanoparticles as Antimicrobial Agents, Front. Pharmacol. 13 (2022).

https://doi.org/10.3389/fphar.2022.895616.

[96] M. Guidotti-Takeuchi, L.N. de M. Ribeiro, F.A.L. dos Santos, D.A. Rossi, F. Della Lucia, R.T. de Melo, Essential Oil-Based Nanoparticles as Antimicrobial Agents in the Food Industry, Microorganisms 10 (2022). https://doi.org/10.3390/microorganisms10081504.

[97] S. Mertdinç-Ülküseven, K. Onbasli, E. Çakır, Y. Morova, Ö. Balcı-Çağıran, H.Y. Acar, A. Sennaroğlu, M. Lütfi Öveçoğlu, D. Ağaoğulları, Magnetic core/shell structures: A case study on the synthesis and phototoxicity/cytotoxicity tests of multilayer graphene encapsulated Fe/Fe3C nanoparticles, J. Alloys Compd. 968 (2023). https://doi.org/10.1016/j.jallcom.2023.172145.

[98] C. Chen, J. Ge, Y. Gao, L. Chen, J. Cui, J. Zeng, M. Gao, Ultrasmall superparamagnetic iron oxide nanoparticles: A next generation contrast agent for magnetic resonance imaging, Wiley Interdiscip. Rev. Nanomedicine Nanobiotechnology 14 (2022). https://doi.org/10.1002/wnan.1740.

[99] R. V. Khose, P. Bangde, M.P. Bondarde, P.S. Dhumal, M.A. Bhakare, G. Chakraborty, A.K. Ray, P. Dandekar, S. Some, Waste derived approach towards wealthy fluorescent N-doped graphene quantum dots for cell imaging and H2O2 sensing applications, Spectrochim. Acta - Part A Mol. Biomol. Spectrosc. 266 (2022). https://doi.org/10.1016/j.saa.2021.120453.

[100] J.R. Melamed, S.S. Yerneni, M.L. Arral, S.T. LoPresti, N. Chaudhary, A. Sehrawat, H. Muramatsu, M.G. Alameh, N. Pardi, D. Weissman, G.K. Gittes, K.A. Whitehead, Ionizable lipid nanoparticles deliver mRNA to pancreatic β cells via macrophage-mediated gene transfer, Sci. Adv. 9 (2023). https://doi.org/10.1126/sciadv.ade1444.

[101] A.H. Rather, R.S. Khan, T.U. Wani, M. Rafiq, A.H. Jadhav, P.M. Srinivasappa, A. Abdal-hay, P. Sultan, S. ullah Rather, J. Macossay, F.A. Sheikh, Polyurethane and cellulose acetate micro-nanofibers containing rosemary essential oil, and decorated with silver nanoparticles for wound healing application, Int. J. Biol. Macromol. 226 (2023). https://doi.org/10.1016/j.ijbiomac.2022.12.048.

[102] A.M. Alotaibi, N.B. Alsaleh, A.T. Aljasham, E.A. Tawfik, M.M. Almutairi, M.A. Assiri, M. Alkholief, M.M. Almutairi, Silver Nanoparticle-Based Combinations with Antimicrobial Agents against Antimicrobial-Resistant Clinical Isolates, Antibiotics 11 (2022). https://doi.org/10.3390/antibiotics11091219.

[103] S. Bhattacharya, M.M. Anjum, K.K. Patel, Gemcitabine cationic polymeric nanoparticles against ovarian cancer: formulation, characterization, and targeted drug delivery, Drug Deliv. 29 (2022). https://doi.org/10.1080/10717544.2022.2058645.

[104] E. Dabirian, A. Hajipour, A.A. Mehrizi, C. Karaman, F. Karimi, P. Loke-Show, O. Karaman, Nanoparticles application on fuel production from biological resources: A review, Fuel 331 (2023). https://doi.org/10.1016/j.fuel.2022.125682.

[105] T. Patil, R. Gambhir, A. Vibhute, A.P. Tiwari, Gold Nanoparticles: Synthesis Methods, Functionalization and Biological Applications, J. Clust. Sci. 34 (2023). https://doi.org/10.1007/s10876-022-02287-6.

[106] G. Oberdörster, E. Oberdörster, J. Oberdörster, Nanotoxicology: An emerging discipline evolving from studies of ultrafine particles, Environ. Health Perspect. 113 (2005).

https://doi.org/10.1289/ehp.7339.

[107] A. Pourmand, M. Abdollahi, Current opinion on nanotoxicology, DARU, J. Pharm. Sci. 20 (2012). https://doi.org/10.1186/2008-2231-20-95.

[108] H. Bahadar, F. Maqbool, K. Niaz, M. Abdollahi, Toxicity of nanoparticles and an overview of current experimental models, Iran. Biomed. J. 20 (2016).

[109] R. Khursheed, K. Dua, S. Vishwas, M. Gulati, N.K. Jha, G.M. Aldhafeeri, F.G. Alanazi, B.H. Goh, G. Gupta, K.R. Paudel, P.M. Hansbro, D.K. Chellappan, S.K. Singh, Biomedical applications of metallic nanoparticles in cancer: Current status and future perspectives, Biomed. Pharmacother. 150 (2022). https://doi.org/10.1016/j.biopha.2022.112951.

[110] A.T. Odularu, Metal Nanoparticles: Thermal Decomposition, Biomedicinal Applications to Cancer Treatment, and Future Perspectives, Bioinorg. Chem. Appl. 2018 (2018). https://doi.org/10.1155/2018/9354708.

Nanomaterials in Biological Systems
Materials Research Foundations 185 (2026) 55-72

Materials Research Forum LLC
https://doi.org/10.21741/9781644903858-3

Chapter 3

Nanoparticle-Based Approaches for Targeted Drug- Delivery

Jasvir Kaur[1,2,*] Prabhat K. Singh[1,3,*]

[1]Radiation & Photochemistry Division, Bhabha Atomic Research Centre, Mumbai 400 085, India

[2]University Institute of Biotechnology, Chandigarh University, Panjab 140 413, India

[3]Homi Bhabha National Institute, Anushaktinagar, Mumbai-400085, India

kaurjasvir8787@gmail.com, prabhatk@barc.gov.in, prabhatsingh988@gmail.com

Abstract

Nanotechnology, which harnesses the potential of nanosized objects, has ushered in a new era in therapeutic drug delivery. Among the pantheons of nanoscale tools, nanoparticles stand out because of their versatility and adaptability, making them especially promising for biomedical applications. These nanocarriers can be meticulously engineered by taking advantage of their modifiable attributes, such as size, shape, surface chemistry, and solubility. This allows them to deliver a diverse range of drugs to specific locations in the body with enhanced precision. One of the most significant advantages of nanoparticle-based drug delivery systems is their ability to address the longstanding challenges posed by physiological barriers, most notably the blood-brain barrier. By navigating these barriers, nanoparticles present an unparalleled opportunity to treat complex diseases that were previously inaccessible. Their biocompatibility and stability further enhance their suitability for *in vivo* applications. Particularly in the field of oncology, these attributes offer transformative approaches to cancer therapy, ensuring that therapeutic agents reach the tumor sites effectively. This chapter provides an in-depth overview of nanoparticle-based drug delivery platforms, emphasizing their potential for reshaping targeted drug delivery, especially in the context of cancer treatment.

Keywords

Nanocarriers, Target Site Drug Delivery, Cancer, Nanoparticles, Quantum Dots

Contents

1.   Introduction

Advancements in nanoscale technologies have facilitated the development of groundbreaking methodologies for disease diagnosis and treatment. This is highlighted by the exponential growth in the field of nanomaterials, which comprises a plethora of entities, such as nanoparticles, functionalized carbon nanotubes, self-assembled polymeric nano-constructs, and nanosized silicon chips, all of which are designed for drug delivery[1,2]. For example, functionalized carbon nanotubes are cylindrical nanostructures that originate from graphene sheets. They possess high strength and thermal stability, which renders them of considerable interest in the biomedical sector. The term "functionalization" refers to chemical modifications on their surface that enhance solubility biocompatibility and enable the binding of specific therapeutic molecules [1]. Self-assembled polymeric nano-constructs result from the spontaneous organization of individual polymer molecules. These nano-constructs can encapsulate and release drugs in a controlled manner, positioning them as promising candidates for drug delivery mechanisms [2]. Although each category of nanomaterials has its own advantages, nanoparticles have garnered substantial attention in the context of drug delivery. Their appeal lies in their modifiability, encompassing aspects such as size, shape, and surface chemistry, tailored to specific biomedical needs [2]. Additionally, the synthesis of biodegradable nanoparticles has witnessed an upward trend, emphasizing their potential for controlled drug release to specific sites, thereby reducing systemic exposure [1,2].

Targeted drug delivery, especially in conditions such as cancer, accentuates the significance of nanoparticles. Traditional drug delivery methods often lead to systemic distribution, which, while ensuring that the therapeutic reaches the diseased site, also exposes healthy tissues to potential side effects. In conditions such as cancer, where precision is paramount, such a systemic distribution can be suboptimal. Owing to their inherent properties, nanoparticles can leverage the enhanced permeability and retention (EPR) effect, a characteristic feature of tumor tissues, ensuring passive targeting [1]. Furthermore, conjugating nanoparticles with specific ligands or molecules can actively target receptors predominantly expressed on cancerous cells, ensuring a precise therapeutic impact [2].

Materials Research Forum LLC
https://doi.org/10.21741/9781644903858-3

In light of these developments, this chapter provides an in-depth exploration of nanoparticle-dependent, targeted drug delivery. Although the broader domain of nanomaterials and their implications in drug delivery have been studied extensively [1–7], our discourse will concentrate on nanoparticle-centric methodologies. The primary focus will be on their applications in cancer therapy and elucidating mechanisms, advantages, challenges, and prospects.

2. Nanoparticles (NPs)

Nanoparticles are colloidal structures that typically vary in size from 10 to 1000 nm [8,9]. For drug delivery applications, nanoparticles larger than 200 nm are generally not preferred, primarily because their increased size can hinder optimal tissue penetration and cellular uptake. On the other hand, smaller nanoparticles can navigate the body's complex microenvironments more effectively by optimizing therapeutic delivery. In recent years, nanoparticles have garnered substantial attention in drug delivery. This interest stems from their intrinsic advantages. First, their surfaces can be easily modified to cater to specific therapeutic needs, enhancing the target specificity. Second, their inherent stability in biological systems ensures that they retain their structural integrity, preventing the premature release or degradation of the encapsulated drug.

One of the pivotal mechanisms by which nanoparticles enhance drug delivery is the "enhanced permeability and retention (EPR) effect." In simplistic terms, the EPR effect refers to the propensity of certain tissues, especially tumor tissues, to allow nanoparticles to accumulate more than larger entities owing to their leaky vasculature (Figure 1). However, it is crucial to note that while the vasculature is leaky, allowing nanoparticles in it, this unique characteristic ensures that nanocarriers can provide sustained and localized drug release to target sites.

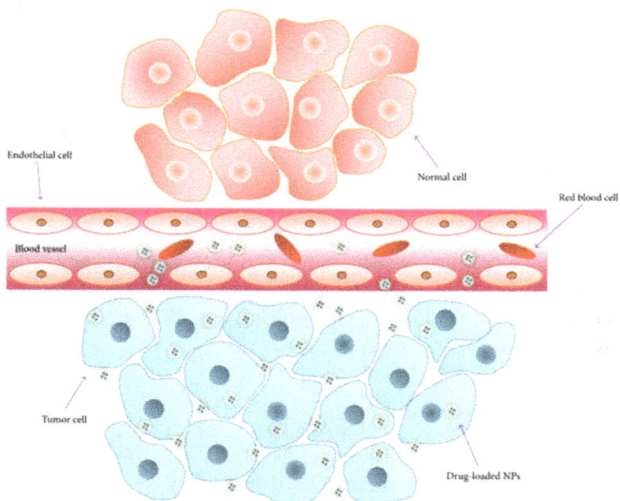

Figure 1. Pictorial representation of enhanced permeation and retention effects (EPR). Adapted with the permission of Yu et al. 2016 [11].

When considering drug encapsulation, encapsulated drug carriers have a distinct edge over their nonencapsulated counterparts. Encapsulation safeguards the therapeutic agent from degradation by external agents such as proteases and nucleases, enhancing the stability and efficacy of the drug. In contrast, nonencapsulated carriers might expose the drug to the external environment, leading to reduced therapeutic effectiveness. Furthermore, the diminutive size of nanoparticles and their potential for biodegradation ensures optimal uptake across various physiological barriers, allowing for sustained drug release within intracellular sites [10]. For the successful development of a nanocarrier-based drug delivery system, it is imperative to consider the defining attributes of nanoparticles, as shown in Figure 2.

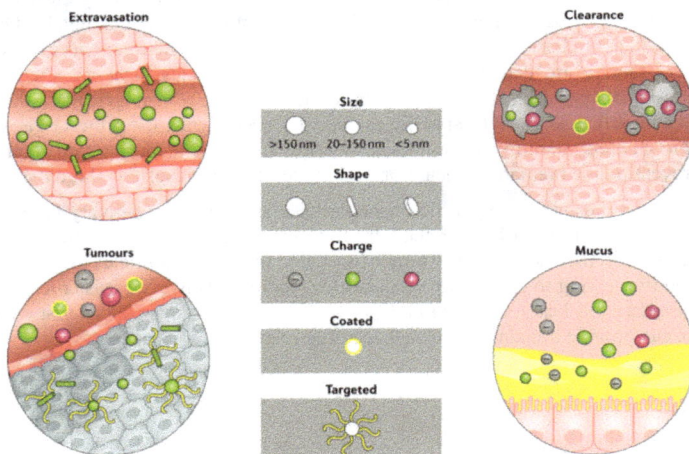

Figure 2: The figure displaying Various characteristic features of nanoparticles that impact their distribution in the circulation and their interaction with local environmental barriers, such as tumors or mucosal layers. Adapted from Mitchell et al. 2021 [5].

2.1   Size and shape of NPs:

The size and shape of nanoparticles (NPs) play pivotal roles in determining the efficiency and effectiveness of drug delivery systems. These parameters intricately influence a number of properties associated with drug delivery, ranging from drug distribution at the target site to its eventual biological fate.

Size: The size of NPs dictates several characteristics.

Drug Delivery and Distribution: The ability of NPs to reach and distribute within a target site is profoundly influenced by their size. For instance, the blood-brain barrier in brain tumors, known for its restrictive permeability, can be successfully traversed by smaller, specifically engineered nanoparticles like tween 80 coated NPs [12].

Cellular Uptake: The efficiency of cellular uptake was inversely related to the NP size. Larger NPs often face challenges in being internalized by cells, whereas smaller NPs are more readily taken up, facilitating more effective drug delivery [13–15].

Drug Release: The drug release rate from NPs is contingent on their surface-area-to-volume ratio. Smaller nanoparticles with a higher ratio tend to release encapsulated drugs more rapidly. This phenomenon has been observed in chemotherapy, where rapid drug release can induce a highly cytotoxic effect on cancer cells [14].

Stability: Stability concerns, especially during storage, transportation, or dispersion, are intrinsically linked to the size of NPs. Smaller NPs may exhibit a higher propensity to aggregate, affecting the overall efficacy of the drug delivery system [16,17].

Shape: While size is undeniably crucial, the shape of NPs also significantly influences drug delivery dynamics.

Circulation Dynamics: Shape determines how NPs navigate through the circulatory system. For example, larger spherical NPs have circulated more smoothly than rod-shaped nanoparticles, which may face challenges in certain vascular environments [5]. This observation directly affects conditions such as cardiovascular diseases, in which ensuring nanoparticle circulation is paramount for therapeutic success.

Tissue Penetration: The shape of the NPs can affect their ability to penetrate tissues. Research in tumor therapy has shown that certain shaped nanoparticles can penetrate tumor masses more effectively, reach deeper cancerous cells, and ensure a more comprehensive therapeutic impact.

Incorporating these considerations during the design and synthesis of NPs can significantly enhance the therapeutic outcomes of NP-based drug delivery systems. By optimizing the size and shape of NPs, researchers can tailor them to specific applications, ensuring maximum efficacy and minimal side effects.

2.2   Surface chemistry

The surface chemistry of drug delivery carriers greatly influences their interactions within a biological milieu. One of the primary challenges these carriers face is their potential recognition and subsequent clearance by the immune system, which could severely compromise their therapeutic efficacy.

Immune Evasion: The size and hydrophobic nature of nanoparticles (NPs) are critical determinants in evading immune surveillance. Their surfaces are often modified to imbue hydrophilic character, which can circumvent rapid clearance mechanisms. This is achieved using a variety of agents.

Polymers and Copolymers: Polyethylene glycol (PEG) is a frequently employed polymer that imparts stealth properties to NPs, allowing them to evade immune recognition and prolong their circulation time in the bloodstream. [18] In real-world applications, PEGylated liposomes have been used to deliver chemotherapeutic agents, providing enhanced therapeutic outcomes and reduced systemic toxicity.

Surfactants: Agents such as Tween 80 and poly-oxamer have hydrophilic groups that can confer favorable surface chemistry to NPs, enhancing their biocompatibility and reducing non-specific interactions with proteins and cells [19]. For example, Tween 80 coated nanoparticles have demonstrated increased brain uptake, making them promising neurotherapeutic candidates.

Charge Considerations: In addition to hydrophilicity, the charge on the surface of the NPs is another pivotal factor. Uncoated nanoparticles with a positive charge are more prone to rapid clearance by the immune system, likely due to their interaction with negatively charged cellular membranes or serum proteins (Figure 2). [5] In practical scenarios, cationic lipid nanoparticles have been employed in gene therapy; however, their positive charge necessitates careful design to mitigate potential cytotoxicity and immune clearance.

Opsonization and Phagocytosis: The surface properties of NPs are intimately linked to their susceptibility to opsonization, wherein serum proteins attach to their surfaces, marking them for clearance by the mononuclear phagocyte system (MPS). Successful alteration of the NP surface chemistry can inhibit this process, ensuring prolonged circulation and effective drug delivery. Real-world applications can be observed in the case of specific antibiotic delivery systems, where surface-modified NPs evade rapid clearance, allowing for the effective targeting of intracellular pathogens.

In conclusion, meticulous tuning of surface chemistry is paramount for optimizing the in vivo behavior of NPs. By understanding and leveraging these principles, we can achieve superior drug delivery outcomes, minimize off-target effects, and maximize therapeutic efficacy.

2.3 Loading, release, and delivery of the drug

A proficient drug delivery system employing nanoparticles (NPs) requires a capacious storage mechanism that ensures significant drug loading via the incorporation method or adsorption/desorption processes. Drug accumulation within nanocarriers is influenced by factors such as the solubility of the drug, the isoelectric point of the macromolecules, and ionic interactions between the drug and nanocarrier matrix [1,20,21].

After evasion of the immune system's surveillance, an imperative step is the efficient release of the drug at the target site. This release process is contingent upon variables such as drug solubility, biodegradation rate of the nanocarrier matrix, drug diffusion through the matrix, and desorption in scenarios involving surface-adsorbed drugs. An illustrative example is that of drugs evenly distributed within nanocarrier systems, such as nanospheres; their release can be mediated either by matrix erosion or drug diffusion. The prevailing diffusion or erosion dynamics govern the pace of drug release. Occasionally, a rapid 'burst' release is noted for drugs loosely bound or adsorbed on the nanocarrier surface. In contrast, a sustained release pattern is typically observed when drugs are introduced via the incorporation method [22]. In 2023, Mitchell et al. meticulously categorized various nanoparticle classes frequently used in drug loading and delivery, emphasizing their defining attributes[5] (Figure 3).

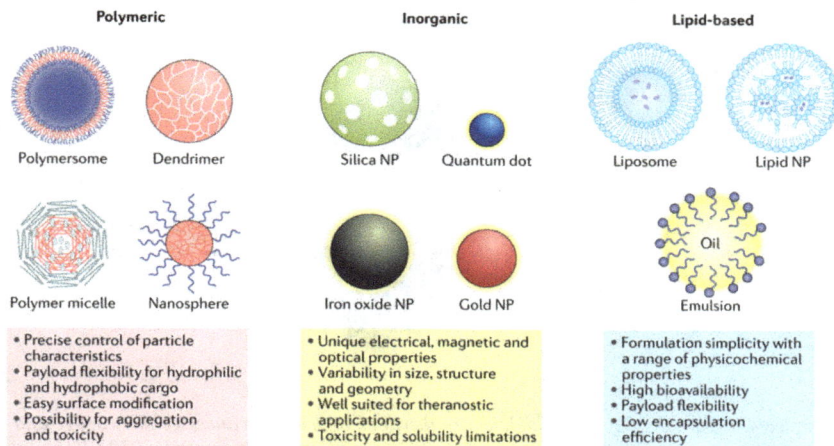

Figure 3. Classes of nanoparticles play an important role in drug loading, disease, and patient response. Adapted from Mitchell et al.. al. 2021 [5].

2.4   Target site drug delivery

In targeted drug delivery, drugs are engineered to act specifically on designated locations within the body. The two predominant strategies in this domain are as follows:

a.) Active Targeting: Here, the drug delivery system or the therapeutic agent itself is conjugated with specific molecules that exhibit affinity to ligands present on cells at the delivery site. For instance, antibodies targeting specific tumor markers can be conjugated to nanoparticles to ensure direct delivery to cancer cells, enhancing therapeutic efficacy and minimizing off-target effects [[23], Figure 4].

b.) Passive Targeting: In strategy, drugs are encased within nanocarriers or tethered to macromolecules, which then passively navigate to the target site. A notable methodology involves the direct infusion of nanocarriers into the target organ using catheters [10,24]. In particular, liposomes have garnered significant attention as drug carriers. They facilitate drug delivery via contact-facilitated mechanisms, providing a sustained-release profile. An exemplary application involves formulating nanoparticles to traverse biological barriers, such as the blood-brain barrier, delivering therapeutics to otherwise challenging sites [2,5,11,26,27]

Comparative Insights: Active vs. Passive Targeting

Active targeting while ensuring precision might be cost-intensive because of the requirement of specific ligands or molecules. It also carries the risk of potential off-target interactions if the targeted receptor is present in non-target cells. Passive targeting, on the other hand, typically incurs lower costs and leverages the inherent physiological properties for drug delivery. However, it might sometimes lack intrinsic specificity to active targeting, which can affect therapeutic outcomes [25].

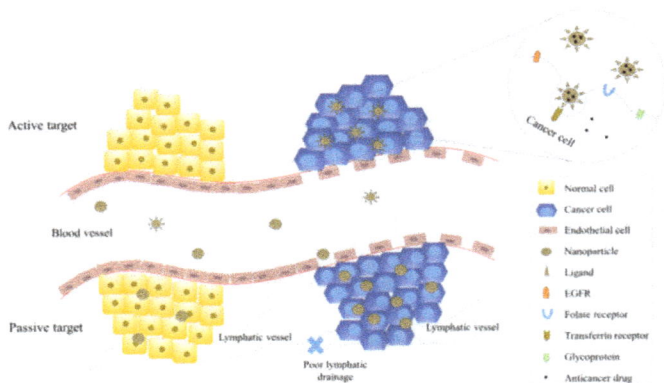

Figure 4. Figure displaying the delivery of therapeutic drugs to cancer cells by active and passive targeting methods. Adapted with permission from Yao et al. 2020 [2].

3. Targeted drug delivery approaches using NPs in cancer cells

The principle behind targeted drug delivery to a specific organ or tissue revolves around the design of vesicular systems that are capable of encasing the drug within a defined compartment. Encapsulation can be achieved through various mechanisms, including entrapment, dissolution, adsorption, encapsulation, or attachment, either within or on the surface of nanosystems, such as nanoparticles, nanocapsules, and nanospheres.

Nanocapsules versus Nanospheres: Based on their structural attributes, a clear distinction exists between nanocapsules and nanospheres. Nanocapsules are vesicular systems in which a drug is enveloped within a cavity and shielded by a polymeric membrane. This ensures a distinct separation between the drug and the surrounding medium, potentially providing a controlled-release mechanism. In contrast, nanospheres are matrix systems in which the drug is uniformly dispersed within the matrix. One of the advantages of nanocapsules is their ability to provide a more controlled release profile, particularly for drugs that are sensitive to the external environment. In contrast, nanospheres, owing to their uniform drug dispersion, might offer a more sustained release mechanism, which is beneficial for prolonged therapeutic effects [28]. Cancer, a complex, multifaceted disease, presents several challenges for drug delivery. The heterogeneous nature of tumors, the presence of a hostile tumor microenvironment, and barriers such as the extracellular matrix often impede effective drug delivery. Nanoparticles, with their customizable properties, have emerged as a promising solution to these challenges by ensuring targeted drug delivery, reduced systemic toxicity, and enhanced therapeutic efficacy. A panoramic overview of the diverse nanoparticles employed in cancer therapy is shown in Figure 5 [2]. In this chapter, we discuss the following NP-based target-specific drug delivery modalities for cancer therapy.

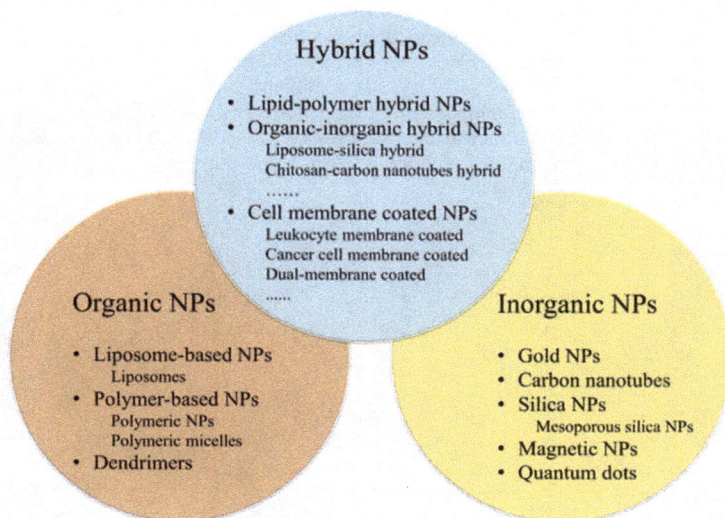

Figure 5. Various classes of nanoparticle-based drug delivery approaches are used for cancer prevention. Adapted with permission from Yao et al. 2020 [2].

3.1 Liposomes and micelles

Liposomes and micelles, as illustrated in Figure 6, are amphiphilic spherical systems that serve as promising drug delivery vehicles. Micelles are characterized by a hydrophobic core formed by hydrophobic molecules, whereas their exterior is composed of hydrophilic molecules. In contrast, liposomes adopt a bilayered arrangement that encloses a hydrophilic core. The choice of where therapeutic drugs are loaded, either within the core or the surrounding shell, largely depends on the hydrophobic or hydrophilic nature of the drug. Once encapsulated, these drugs are protected from potential degradation or rapid clearance by the immune system, thus enhancing their therapeutic efficacy. These synthetic drug-carrying platforms mirror specific natural biological systems in their mode of operation and offer advantages such as biocompatibility and reduced immunogenicity. A quintessential example of the versatility of these systems is the development of a pH- and temperature-responsive drug delivery platform. This platform combines the properties of poly(acrylic acid) (PAA) and heat-sensitive poly(N-isopropylacrylamide) (PNIPAM) [28]. Under specific conditions, pH values below 4.8 for PAA and temperatures exceeding 32°C for PNIPAM, these molecules adopt hydrophobic configurations. This unique behavior facilitates the assembly of co-polymeric micelles, wherein environmental cues, such as pH and temperature, dictate the micelle structure and, consequently, drug release kinetics.

However, why are such responsive systems pivotal for drug delivery? The human body encompasses myriad microenvironments with distinct pH and temperature profiles. For instance, tumor microenvironments often exhibit an acidic pH, whereas inflammation sites may register elevated temperatures. Harnessing these physiological variations, pH- and temperature-responsive systems can be engineered to precisely release drugs where they are most needed, ensuring targeted therapy while minimizing off-target effects. In the provided example, the complex of PAA and doxorubicin (DOX) forms micelles under specific conditions, ensuring the controlled release of the drug in response to pH or temperature shifts. Similarly, nanoparticles conjugated with PNIPAM have been explored for their temperature responsiveness, capitalizing on the properties of the polymer to achieve controlled drug release at designated sites [3].

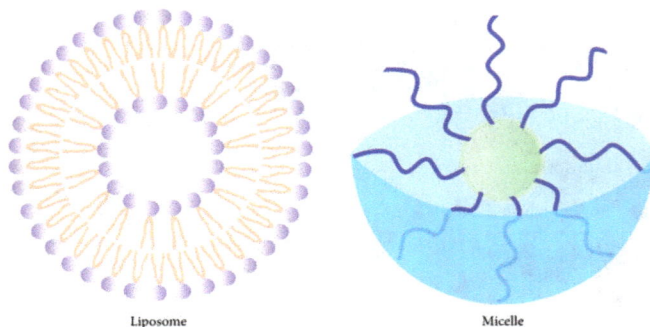

Liposome

Micelle

Figure 6. Schematic representation of liposomes and micelles of the two types of amphiphilic nanoparticles. Adapted with permission from Yu and co-workers 2016 [11].

3.2   Quantum Dots (QDs)

Although lung cancer remains a predominant health challenge, with statistics from the American Cancer Society designating it as the leading cause of death in the Western world [29], the advent of Quantum Dots (QDs) and semiconductor crystals in cancer therapeutics is a beacon of hope. These nanostructures, endowed with unique chemical and physical attributes, such as diminutive size and modifiable surface chemistry, have shown considerable promise in enhancing solubility and biocompatibility [30–32]. Their versatility does not end in lung cancer. Quantum Dots have been explored in the diagnosis and treatment of other malignancies, including breast and pancreatic cancers, and even in imaging applications for neurological diseases. [27,33]

One of the standout therapeutic approaches involves the functionalization of QD centers with specific ligands, such as folic acid. This strategy colloquially termed the 'Trojan horse', facilitates targeted drug delivery. The mechanism illustrated in Figure 7 proceeds as follows.

a) Drug-conjugated folic acid molecules recognize and bind to cell-surface receptors.

b) This binding triggers receptor-mediated endocytosis and internalization of folic acid conjugates.

c) In the cellular environment, endosomal digestion processes disintegrate the conjugates, liberating the encapsulated drugs at their intended target and recycling receptors to the cell surface.

Gu et al. exemplify this method by employing Ag–In–Zn–S QDs modified with a folic acid-doxorubicin conjugate, capitalizing on the ability of doxorubicin to inhibit topoisomerase and thereby curtail cancer proliferation [34]. In a parallel approach, the human epidermal growth factor receptor (HER2) has been targeted using HER2 anti-monoclonal antibody (MAB)-conjugated QDs (QD-MAB). Post-injection, the QD-MAB complex latches onto HER2 receptors present in tumor cells, facilitating their internalization and subsequent delivery to the perinuclear region of the cell [35].

Although the promise of QDs is undeniable, it is essential to temper optimism cautiously. Quantum Dots, especially those composed of heavy metals, raise concerns regarding their potential cytotoxicity and long-term accumulation in biological systems. Moreover, their stability in physiological environments and their potential for off-target effects warrant further research. As with any emerging technology, a comprehensive understanding of QDs' therapeutic potential of QDs must be juxtaposed with their limitations to harness their capabilities.

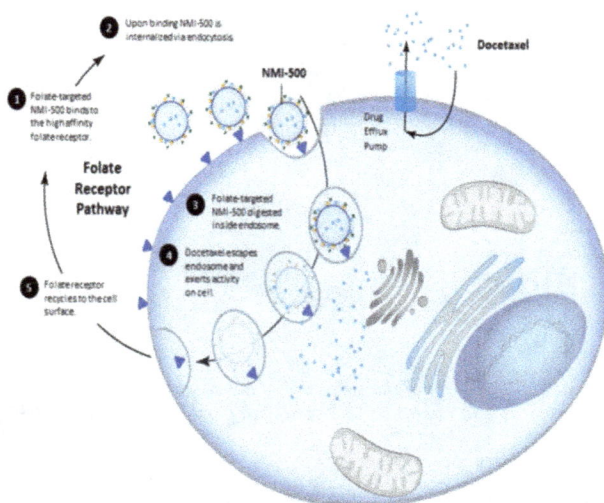

Figure 7. Pictorial representation of receptor-mediated endocytosis of NPs (MI-500) loaded with drug docetaxel internalized using folate receptors expressed on the surface of cancer cells. Adapted with permission from Patel 2018 [27].

3.3   Rotaxanes and Dendrimers

Rotaxanes are fascinating molecular architectures wherein a cyclic molecule encircles a dumbbell-shaped entity. Although this cyclic molecule enjoys rotational freedom around the dumbbell, it cannot dissociate from it (Figure 8A) [36]. A notable application of this structure is evident in a study where cyclodextrin, a cyclic molecule, was conjugated to polyethylene glycol (PEG), which in turn was linked to cinnamic acid. This design aimed to augment the interspace between cyclodextrin rings, facilitating the binding of therapeutic agents, such as doxorubicin and methotrexate [36,37].

On the other hand, dendrimers are intricate nanostructures that evoke images of branching blood vessels or tree limbs. This branching design gives them an expansive surface area and ensures a uniform distribution, as depicted in Figure 8B [38]. However, why is this branching important, particularly for drug delivery? The answer lies in the dendrimer mimicry of the vascular system. Through their intricate branching, blood vessels ensure efficient transport and delivery of nutrients and oxygen to every cellular nook and cranny. Similarly, by mimicking this branching, dendrimers can transport therapeutic agents throughout their structure, ensuring efficient drug encapsulation, protection, and subsequent release at the target site.

To exploit these benefits, dendrimers were fabricated using polyethylene glycol (PEG) and dendritic cholic acid (CA) molecules [39]. These dendritic platforms have demonstrated potential in the field of oncology for effectively delivering anticancer drugs to malignant cells [40].

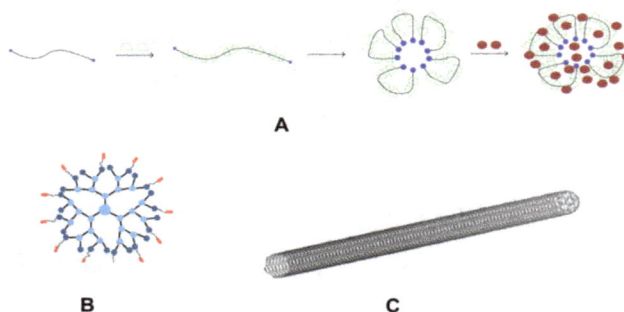

Figure 8. Schematic illustration of the structure of novel nanoparticles A. Polyrotaxane B. Dendrimer C. Carbon nanotube. Adapted with permission from Yu 2016 [11].

3.4   Nanotubes and hydrogels

Carbon nanotubes are essentially graphene sheets that are artfully rolled to form tubular structures, retaining the intrinsic properties of graphene (as depicted in Figure 8C). One of the remarkable features of carbon nanotubes in the realm of drug delivery is their responsiveness to environmental cues, such as heat or pH, enabling controlled drug release at target sites [41,42]. Single-walled carbon nanotubes (SWCNT) have been used to release antitumor drugs when conjugated with antitumor monoclonal antibodies. Interestingly, some evolved versions of these nanotubes encompass not only antitumor drug antibodies but also nanoparticles known as fullerenes (C60), further enhancing the precision of drug delivery [43,44].

Fullerene-based drug delivery systems offer several advantages.

1. Their capability to simultaneously deliver multiple drugs is due to the plurality of fullerenes that can adhere to the hydrophobic surfaces of antibody molecules.

2. The absence of a covalent bond between fullerenes and antibodies is not just a structural nuance; it is pivotal for drug delivery. Without such a bond, the inherent targeting specificity of the

Nanomaterials in Biological Systems   Materials Research Forum LLC
Materials Research Foundations 185 (2026) 55-72   https://doi.org/10.21741/9781644903858-3

antibody remains uncompromised, ensuring that the drug reaches the designated tumor cells without any aberrations.

However, although fullerenes offer these advantages, they are not without challenges. Their potential for aggregation, cytotoxicity, and the need for functionalization to enhance their solubility in physiological environments warrant further investigation. In addition, nanocarriers have been developed using proprietary technologies. This technology employs hydrophobic polysaccharides to form the core of these carriers, thereby facilitating the encapsulation of therapeutic molecules. The encapsulation of curcumin, a spice renowned for its purported anticancer properties, within polymeric nanoparticles accentuates target-site drug delivery [45].

3.5   Polymersomes

Polymerosomes are vesicles constructed from block copolymers and are characterized by their robust, thick membranes and hollow interior. This unique architecture allows them to encapsulate and subsequently deliver therapeutic agents to target sites. One of the pivotal mechanisms enabling drug release from polymerosomes is pH triggering, which is orchestrated by the formation of endolysosomes in tumor cells. For instance, potent anticancer drugs, such as paclitaxel and doxorubicin, have been successfully enveloped within polymerosomes, directing them precisely to malignancies [46]. Several distinctions emerge when polymerosomes are juxtaposed with other nanoparticle-based drug delivery systems.

Efficiency: Due to their thicker membranes, polymerosomes can often provide more controlled and sustained drug release than other nanoparticle systems. This ensures that the drug is released over an extended period of time, potentially reducing the frequency of dosing and maximizing therapeutic efficacy.

Cost: The synthesis of polymerosomes might be more intricate and, hence, slightly costlier than that of simpler nanoparticle systems. However, their potential for enhanced therapeutic efficacy and reduced dosing frequency can offset these initial cost implications in the long term.

Potential Side Effects: As Similar to any drug delivery system, polymerosomes encounter a set of challenges. Biocompatibility and biodegradability of polymers are of paramount importance. While polymerosomes generally exhibit good biocompatibility, it is essential to ensure that any degradation products of the copolymers are non-toxic.

In summary, while polymerosomes are a promising avenue in targeted drug delivery, their deployment needs to be context-specific, weighing their advantages against potential challenges and comparing them to other available nanoparticulate systems.

Conclusion

In this chapter, we delved into the transformative potential of nanoparticles in devising target-specific drug delivery systems. Their stability, biocompatibility, ease of functionalization, and broad adaptability have made them indispensable for contemporary therapeutic research. Nanoparticles can be engineered to deliver diverse drugs to designated body regions by manipulating attributes, such as size, shape, charge, and surface properties. Drug delivery strategies can hinge on passive targeting, capitalizing on the enhanced permeability and retention effect, or active targeting, which employs receptor ligand-mediated endocytosis mechanisms. The versatility of nanoparticles is further exemplified by the advent of photothermal and magnetic hybrid nanocarriers, which can be modulated based on the location and timing of the target site. Our discussion spans an array of novel structures that have burgeoned from nanoparticle research, encompassing dendrimers, rotaxanes, quantum dots, liposomes, micelles, nanotubes, hydrogels, and polymerosomes. These entities imbue nanocarriers with the requisite traits for specialized drug delivery to distinct organs or tissues. When juxtaposed with traditional drug delivery paradigms, nanoparticle-based systems offer enhanced solubility, stability, bioavailability, specificity, efficacy, and safety. However, the narrative of nanoparticle-based drug delivery is challenging. Concerns include potential toxicity, unforeseen immunogenic responses, biodistribution dynamics, clearance rates, unwanted accumulation, and the emergence of resistance mechanisms. It is imperative to approach these challenges from a balanced perspective, cognizant of both the potential and pitfalls. The horizon of nanoparticle-based drug delivery is promising, yet it requires more rigorous research. Optimization of the design, synthesis, characterization, evaluation, and regulatory pathways remains paramount, particularly in the context of cancer therapeutics and a spectrum of other biomedical applications. Future prospects in this domain include refining the biocompatibility profiles, enhancing target specificity, and harnessing advanced imaging techniques. Active clinical trials and innovations in pipelines are likely to augment our understanding and applications further. Finally, the intricacy of nanoparticle research emphasizes the importance of interdisciplinary collaboration. Advancing this domain requires cohesive endeavors from chemists, biologists, medical practitioners, and regulatory bodies. Such a confluence of expertise will be instrumental in revolutionizing cancer treatment and enhancing patient outcomes.

References

[1] R. Singh, J.W. Lillard, Nanoparticle-based targeted drug delivery, Experimental and Molecular Pathology. 86 (2009) 215–223. https://doi.org/10.1016/j.yexmp.2008.12.004

[2] Y. Yao, Y. Zhou, L. Liu, Y. Xu, Q. Chen, Y. Wang, S. Wu, Y. Deng, J. Zhang, A. Shao, Nanoparticle-Based Drug Delivery in Cancer Therapy and Its Role in Overcoming Drug Resistance, Frontiers in Molecular Biosciences. 7 (2020). https://www.frontiersin.org/articles/10.3389/fmolb.2020.00193 (accessed July 28, 2023).

[3] A. Rahikkala, V. Aseyev, H. Tenhu, E.I. Kauppinen, J. Raula, Thermoresponsive Nanoparticles of Self-Assembled Block Copolymers as Potential Carriers for Drug Delivery and Diagnostics, Biomacromolecules. 16 (2015) 2750–2756. https://doi.org/10.1021/acs.biomac.5b00690

[4] V. Chandrakala, V. Aruna, G. Angajala, Review on metal nanoparticles as nanocarriers: current challenges and perspectives in drug delivery systems, Emergent Mater. 5 (2022) 1593–1615. https://doi.org/10.1007/s42247-021-00335-x

[5] M.J. Mitchell, M.M. Billingsley, R.M. Haley, M.E. Wechsler, N.A. Peppas, R. Langer, Engineering precision nanoparticles for drug delivery, Nat Rev Drug Discov. 20 (2021) 101–124. https://doi.org/10.1038/s41573-020-0090-8

[6] J.N. Cruz, S. Muzammil, A. Ashraf, M.U. Ijaz, M.H. Siddique, R. Abbas, M. Sadia, Saba, S. Hayat, R.R. Lima, A review on mycogenic metallic nanoparticles and their potential role as antioxidant, antibiofilm and quorum quenching agents, Heliyon. 10 (2024). https://doi.org/10.1016/j.heliyon.2024.e29500

[7] R. Liu, C. Luo, Z. Pang, J. Zhang, S. Ruan, M. Wu, L. Wang, T. Sun, N. Li, L. Han, J. Shi, Y. Huang, W. Guo, S. Peng, W. Zhou, H. Gao, Advances of nanoparticles as drug delivery systems for disease diagnosis and treatment, Chinese Chemical Letters. 34 (2023) 107518. https://doi.org/10.1016/j.cclet.2022.05.032

[8] J. Kaur, P.K. Singh, Nanomaterial based advancement in the inorganic pyrophosphate detection methods in the last decade: A review, TrAC Trends in Analytical Chemistry. 146 (2022) 116483. https://doi.org/10.1016/j.trac.2021.116483

[9] J. Swarbrick, Encyclopedia of Pharmaceutical Technology: Volume 6, 3rd ed., CRC Press, Boca Raton, 2015. https://doi.org/10.1201/b19309

[10] H. Maeda, The enhanced permeability and retention (EPR) effect in tumor vasculature: the key role of tumor-selective macromolecular drug targeting, Adv Enzyme Regul. 41 (2001) 189–207. https://doi.org/10.1016/s0065-2571(00)00013-3

[11] X. Yu, I. Trase, M. Ren, K. Duval, X. Guo, Z. Chen, Design of Nanoparticle-Based Carriers for Targeted Drug Delivery, Journal of Nanomaterials. 2016 (2016) e1087250. https://doi.org/10.1155/2016/1087250

[12] J. Kreuter, P. Ramge, V. Petrov, S. Hamm, S.E. Gelperina, B. Engelhardt, R. Alyautdin, H. von Briesen, D.J. Begley, Direct evidence that polysorbate-80-coated poly(butylcyanoacrylate) nanoparticles deliver drugs to the CNS via specific mechanisms requiring prior binding of drug to the nanoparticles, Pharm Res. 20 (2003) 409–416. https://doi.org/10.1023/a:1022604120952

[13] D. Mp, L. V, W. E, L. Rj, A. Gl, The mechanism of uptake of biodegradable microparticles in Caco-2 cells is size dependent, Pharmaceutical Research. 14 (1997). https://doi.org/10.1023/a:1012126301290

[14] H.M. Redhead, S.S. Davis, L. Illum, Drug delivery in poly(lactide-co-glycolide) nanoparticles surface modified with poloxamer 407 and poloxamine 908: in vitro characterisation and in vivo evaluation, J Control Release. 70 (2001) 353–363. https://doi.org/10.1016/s0168-3659(00)00367-9

[15] M.H. Sarfraz, M. Zubair, B. Aslam, A. Ashraf, M.H. Siddique, S. Hayat, J.N. Cruz, S. Muzammil, M. Khurshid, M.F. Sarfraz, A. Hashem, T.M. Dawoud, G.D. Avila-Quezada, E.F. Abd_Allah, Comparative analysis of phyto-fabricated chitosan, copper oxide, and chitosan-based CuO nanoparticles: antibacterial potential against Acinetobacter baumannii isolates and anticancer activity against HepG2 cell lines, Front. Microbiol. 14 (2023). https://doi.org/10.3389/fmicb.2023.1188743

[16] J. Panyam, M.M. Dali, S.K. Sahoo, W. Ma, S.S. Chakravarthi, G.L. Amidon, R.J. Levy, V. Labhasetwar, Polymer degradation and in vitro release of a model protein from poly(D,L-lactide-co-glycolide) nano- and microparticles, J Control Release. 92 (2003) 173–187. https://doi.org/10.1016/s0168-3659(03)00328-6

[17] M. Dunne, O.I. Corrigan, Z. Ramtoola, Influence of particle size and dissolution conditions on the degradation properties of polylactide-co-glycolide particles, Biomaterials. 21 (2000) 1659–1668. https://doi.org/10.1016/S0142-9612(00)00040-5

[18] D. Bhadra, S. Bhadra, P. Jain, N.K. Jain, Pegnology: a review of PEG-ylated systems, Pharmazie. 57 (2002) 5–29.

[19] J.-C. Olivier, Drug transport to brain with targeted nanoparticles, NeuroRx. 2 (2005) 108–119. https://doi.org/10.1602/neurorx.2.1.108

[20] T. Govender, T. Riley, T. Ehtezazi, M.C. Garnett, S. Stolnik, L. Illum, S.S. Davis, Defining the drug incorporation properties of PLA–PEG nanoparticles, International Journal of Pharmaceutics. 199 (2000) 95–110. https://doi.org/10.1016/S0378-5173(00)00375-6

[21] S. Muzammil, J. Neves Cruz, R. Mumtaz, I. Rasul, S. Hayat, M.A. Khan, A.M. Khan, M.U. Ijaz, R.R. Lima, M. Zubair, Effects of Drying Temperature and Solvents on In Vitro Diabetic Wound Healing Potential of Moringa oleifera Leaf Extracts, Molecules. 28 (2023). https://doi.org/10.3390/molecules28020710

[22] Y. Chen, R.K. McCulloch, B.N. Gray, Synthesis of albumin-dextran sulfate microspheres possessing favourable loading and release characteristics for the anticancer drug doxorubicin, Journal of Controlled Release. 31 (1994) 49–54. https://doi.org/10.1016/0168-3659(94)90250-X

[23] A. Lamprecht, N. Ubrich, H. Yamamoto, U. Schäfer, H. Takeuchi, P. Maincent, Y. Kawashima, C.M. Lehr, Biodegradable nanoparticles for targeted drug delivery in treatment of inflammatory bowel disease, J Pharmacol Exp Ther. 299 (2001) 775–781.

[24] SK Sahoo, T. Sawa, J. Fang, S. Tanaka, Y. Miyamoto, T. Akaike, H. Maeda, Pegylated zinc protoporphyrin: a water-soluble heme oxygenase inhibitor with tumor-targeting capacity, Bioconjug Chem. 13 (2002) 1031–1038. https://doi.org/10.1021/bc020010k

[25] L.A. Guzman, V. Labhasetwar, C. Song, Y. Jang, A.M. Lincoff, R. Levy, E.J. Topol, Local intraluminal infusion of biodegradable polymeric nanoparticles. A novel approach for prolonged drug delivery after balloon angioplasty, circulation. 94 (1996) 1441–1448. https://doi.org/10.1161/01.cir.94.6.1441

[26] R.S. Fisher, J. Ho, Potential new methods for antiepileptic drug delivery, CNS Drugs. 16 (2002) 579–593. https://doi.org/10.2165/00023210-200216090-00001

[27] N.R. Patel, A. Piroyan, S. Ganta, A.B. Morse, K.M. Candiloro, A.L. Solon, A.H. Nack, C.A. Galati, C. Bora, M.A. Maglaty, S.W. O'Brien, S. Litwin, B. Davis, D.C. Connolly, T.P. Coleman, In Vitro and In Vivo evaluation of a novel folate-targeted theranostic nanoemulsion of docetaxel for imaging and improved anticancer activity against ovarian cancers, Cancer Biology & Therapy. 19 (2018) 554–564. https://doi.org/10.1080/15384047.2017.1395118

[28] G. Li, S. Song, L. Guo, S. Ma, Self-assembly of thermo- and pH-responsive poly(acrylic acid)-b-poly(N-isopropylacrylamide) micelles for drug delivery, Journal of Polymer Science Part A: Polymer Chemistry. 46 (2008) 5028–5035. https://doi.org/10.1002/pola.22831

[29] Lung Cancer Statistics | How Common is Lung Cancer?, (n.d.). https://www.cancer.org/cancer/types/lung-cancer/about/key-statistics.html (accessed September 15, 2023).

[30] S.-W. Ha, D. Weiss, M.N. Weitzmann, G.R. Beck, Chapter 4 - Applications of silica-based nanomaterials in dental and skeletal biology, in: K. Subramani, W. Ahmed (Eds.),

Nanobiomaterials in Clinical Dentistry (Second Edition), Elsevier, 2019: pp. 77–112.
https://doi.org/10.1016/B978-0-12-815886-9.00004-8

[31] I.N. de F. Ramos, M.F. da Silva, J.M.S. Lopes, J.N. Cruz, F.S. Alves, J. de A.R. do Rego,
M.L. da Costa, P.P. de Assumpção, D. do S. Barros Brasil, A.S. Khayat, Extraction,
Characterization, and Evaluation of the Cytotoxic Activity of Piperine in Its Isolated form and
in Combination with Chemotherapeutics against Gastric Cancer, Molecules. 28 (2023).
https://doi.org/10.3390/molecules28145587

[32] K. David Wegner, N. Hildebrandt, Quantum dots: bright and versatile in vitro and in vivo
fluorescence imaging biosensors, Chemical Society Reviews. 44 (2015) 4792–4834.
https://doi.org/10.1039/C4CS00532E

[33] A. Sosnik, Chapter 1 - From the "Magic Bullet" to Advanced Nanomaterials for Active
Targeting in Diagnostics and Therapeutics, in: B. Sarmento, J. das Neves (Eds.), Biomedical
Applications of Functionalized Nanomaterials, Elsevier, 2018: pp. 1–32.
https://doi.org/10.1016/B978-0-323-50878-0.00001-X

[34] Y.-J. Gu, J. Cheng, C.W.-Y. Man, W.-T. Wong, S.H. Cheng, Gold-doxorubicin
nanoconjugates for overcoming multidrug resistance, Nanomedicine: Nanotechnology,
Biology and Medicine. 8 (2012) 204–211. https://doi.org/10.1016/j.nano.2011.06.005

[35] H. Tada, H. Higuchi, T.M. Wanatabe, N. Ohuchi, In vivo real-time tracking of single
quantum dots conjugated with monoclonal anti-HER2 antibody in tumors of mice, Cancer
Res. 67 (2007) 1138–1144. https://doi.org/10.1158/0008-5472.CAN-06-1185

[36] R. Liu, Y. Lai, B. He, Y. Li, G. Wang, S. Chang, Z. Gu, Supramolecular nanoparticles
generated by the self-assembly of polyrotaxanes for antitumor drug delivery, Int J
Nanomedicine. 7 (2012) 5249–5258. https://doi.org/10.2147/IJN.S33649

[37] L. Zhang, T. Su, B. He, Z. Gu, Self-assembly Polyrotaxanes Nanoparticles as Carriers for
Anticancer Drug Methotrexate Delivery, Nano-Micro Lett. 6 (2014) 108–115.
https://doi.org/10.1007/BF03353774

[38] S. Svenson, D.A. Tomalia, Dendrimers in biomedical applications—reflections on the field,
Advanced Drug Delivery Reviews. 64 (2012) 102–115.
https://doi.org/10.1016/j.addr.2012.09.030

[39] Y. Li, K. Xiao, J. Luo, W. Xiao, J.S. Lee, A.M. Gonik, J. Kato, T.A. Dong, K.S. Lam, Well-
defined, reversible disulfide cross-linked micelles for on-demand paclitaxel delivery,
Biomaterials. 32 (2011) 6633–6645. https://doi.org/10.1016/j.biomaterials.2011.05.050

[40] AK Patri, A. Myc, J. Beals, T.P. Thomas, N.H. Bander, J.R. Baker, Synthesis and in vitro
testing of J591 antibody-dendrimer conjugates for targeted prostate cancer therapy, Bioconjug
Chem. 15 (2004) 1174–1181. https://doi.org/10.1021/bc0499127

[41] A.C. Estrada, A.L. Daniel-da-Silva, T. Trindade, Photothermally enhanced drug release by
κ-carrageenan hydrogels reinforced with multi-walled carbon nanotubes, RSC Adv. 3 (2013)
10828. https://doi.org/10.1039/c3ra40662h

[42] H. Zhang, C. Chen, L. Hou, N. Jin, J. Shi, Z. Wang, Y. Liu, Q. Feng, Z. Zhang, Targeting
and hyperthermia of doxorubicin by the delivery of single-walled carbon nanotubes to EC-

109 cells, Journal of Drug Targeting. 21 (2013) 312–319.
https://doi.org/10.3109/1061186X.2012.749880

[43] J.M. Ashcroft, D.A. Tsyboulski, K.B. Hartman, T.Y. Zakharian, J.W. Marks, R.B. Weisman, M.G. Rosenblum, L.J. Wilson, Fullerene (C60) immunoconjugates: interaction of water-soluble C60 derivatives with the murine anti-gp240 melanoma antibody, Chem. Commun. (2006) 3004. https://doi.org/10.1039/b601717g

[44] F.S. Alves, J.N. Cruz, I.N. de Farias Ramos, D.L. do Nascimento Brandão, R.N. Queiroz, G.V. da Silva, G.V. da Silva, M.F. Dolabela, M.L. da Costa, A.S. Khayat, J. de Arimatéia Rodrigues do Rego, D. do Socorro Barros Brasil, Evaluation of Antimicrobial Activity and Cytotoxicity Effects of Extracts of Piper nigrum L. and Piperine, Separations. 10 (2023). https://doi.org/10.3390/separations10010021

[45] S. Bisht, G. Feldmann, S. Soni, R. Ravi, C. Karikar, A. Maitra, A. Maitra, Polymeric nanoparticle-encapsulated curcumin ("nanocurcumin"): a novel strategy for human cancer therapy, J Nanobiotechnol. 5 (2007) 3. https://doi.org/10.1186/1477-3155-5-3

[46] F. Ahmed, R.I. Pakunlu, G. Srinivas, A. Brannan, F. Bates, M.L. Klein, T. Minko, D.E. Discher, Shrinkage of a Rapidly Growing Tumor by Drug-Loaded Polymersomes: pH-Triggered Release through Copolymer Degradation, Mol. Pharmaceutics. 3 (2006) 340–350. https://doi.org/10.1021/mp050103u

Nanomaterials in Biological Systems
Materials Research Foundations 185 (2026) 73-91

Materials Research Forum LLC
https://doi.org/10.21741/9781644903858-4

Chapter 4

Nanoparticle-Based Approaches for Wound Healing

Kajal Bhardwaj[1], Vinay Kant[1*], Dhaval J Kamothi[2], Pooja Kumari[3], Babu Lal Jangir[4] and Munish Ahuja[3]

[1]Department of Veterinary Pharmacology & Toxicology, COVS, Lala Lajpat Rai University of Veterinary & Animal Sciences, Hisar-125004, Haryana, India

[2]Deparment of Veterinary Pharmacology & Toxicology, COVS & A.H., Kamdhenu University,Bhuj, India

[3]Dept of Pharmaceutical Sciences, GJU S&T, Hisar, Haryana, India.

[4]Department of Veterinary Pathology, COVS, Lala Lajpat Rai University of Veterinary & Animal Sciences, Hisar, Haryana, India

drvinaykantluvas@gmail.com drvinaykant@luvas.edu.in

Abstract

Wound healing is a biological process involving sequential biochemical events that restore organ cellular integrity. The wound-care product market has surpassed fifteen billion US dollars. Conventional and newer treatments like topical antiseptics, antimicrobials, hydrogels, foam, alginates, and hydro-fibers exist, yet carry disadvantages. Many compounds including phytochemicals hold promise for wound healing, yet factors like light sensitivity, low solubility, poor bioavailability, and less skin permeability limit their clinical use. Recently, nanotechnology conquered some of these obstacles and enhanced stability, delivery, and effectiveness. This chapter explores overview of wound healing and available conventional therapies, along with nanoparticle-based approaches to treat wounds.

Keywords

Skin, Wound Healing, Conventional Treatments, Nanoparticles, Nano-Based Systems

Contents

Nanomaterials in Biological Systems   Materials Research Forum LLC
Materials Research Foundations 185 (2026) 73-91   https://doi.org/10.21741/9781644903858-4

1.   Introduction to wound

Skin, our body's largest organ and protective barrier, consists of the epidermis, dermis, and hypodermis (fat layer) (1). Epidermis, the skin's outermost layer, maintains homeostasis, defends against external elements, and regulates body moisture with its waterproof barrier. Comprised mainly of keratinocytes, it also houses melanocytes, Langerhans cells, and Merkel cells. This multilayered structure includes basal cell division, migration to upper layers, and the formation of dead surface cells. Nestled between the epidermis and the hypodermis, the dermis plays a pivotal role in the skin's overall vitality. It serves as a stronghold, offering structural integrity while accommodating an impressive array of vital components. It houses blood vessels, nerves, hair follicles, and sweat glands, all cohesively integrated. Composed of connective tissue, including the extraordinary extracellular matrix (ECM), vascular endothelial cells, fibroblasts, adipose glands, sweat glands, hair follicles, blood vessels, and nerve endings, the dermis orchestrates a symphony of support (2). The star performers, fibroblasts, take center stage, releasing collagen and elastin, imbuing the skin with remarkable mechanical strength and delightful elasticity. The hypodermis, the skin's lowest layer, is made up of loose connective tissue, fat cells (which store around half of the body's fat), blood arteries, and nerves. This tissue contains a lot of proteoglycans and glycosaminoglycans, which absorb moisture and give it a mucous-like texture (3). With remarkable regenerative abilities, skin orchestrates a cascade of physiological events to heal injuries and cuts efficiently (4). Skin in natural state is dry and acidic (pH 4 to 6.8) (5).

Any breakage in skin integrity is defined as wound that can be due to physical, chemical, or thermal damage, as well as medical conditions or external factors, causing the loss of cohesion in the epithelium with or without loss of connective tissue (6). This skin damage, which can be caused by burns, trauma, surgeries, or lacerations, impairs its function and necessitates proper wound care to prevent complications and hindered healing. Severe skin injuries, such as burns, have the potential to lead to dehydration, infections, and even death, while skin integrity impairment can vary from surface breaks to deeper damage involving tendons, muscles, organs, or bones (5).

These wounds can be classified based on various criteria, including their type and healing process. They are categorized as open or closed wounds, with close wounds involving damage to soft tissue, small blood vessels, or deep tissue layers, and open wounds encompassing various types such as lacerations, surgical wounds, and more (7). Wounds are further classified as acute or chronic, with acute wounds healing within 8 to 12 weeks, while chronic wounds like burns, pressure ulcers, and leg ulcers have an unpredictable healing timeframe. Additionally, wounds can be classified based on specific pathological conditions such as diabetic wounds, burn wounds, surgical wounds, traumatic wounds, and pressure ulcers (6).

2.   Wound healing mechanism

Generally wound healing comprises of 4 stages viz hemostasis, inflammation, proliferation and remodeling phase. In normal/ acute wound healing theses stages occur as follows-

Immediately following an injury, hemostasis swiftly restores the barrier function as the first stage of wound healing. This process involves vasoconstriction, primary and secondary hemostasis, with platelets and fibrinogen playing crucial roles in stopping bleeding and forming the clot (8). This clot acts as a scaffold, facilitating the migration of immune and skin cells. Platelets within the clot release various proteins, growth factors, and cytokines that trigger the activation and migration of macrophages, neutrophils, fibroblasts, smooth muscle cells, endothelial cells, and circulating bone marrow-derived stem cells to the wound site (9).

Inflammatory phase got triggered within hours after injury, is fueled by bacterial byproducts, cytokines/chemokines, and platelet-derived mediators. This inflammatory response commences immediately upon wounding and typically persists for a duration of 1 to 4 days. Neutrophils are the first to infiltrate the wound site, followed by monocytes/macrophages, eliminating bacteria and clearing debris. Neutrophils release proteases, antimicrobial peptides, reactive oxygen species (ROS), and growth factors, while monocytes transform into pro-inflammatory M1 macrophages. Lymphocytes, including γδ+ and αβ+ T cells, contribute to immune cell, fibroblast, and keratinocyte survival and growth. In the later stage, M2 macrophages, derived from monocytes or switched from M1 macrophages, release factors (VEGF, PDGF, IGF-1, FGFs, TGF-β1 and IL-10) promoting cell migration, proliferation, and matrix formation, enabling angiogenesis and tissue granulation for the transition to the proliferation stage. Following the resolution of inflammation and the transition of monocytes and macrophages into anti-inflammatory M2 macrophages, the proliferation phase commences. This transition is facilitated by the immunomodulatory mediators primarily secreted by MSCs (mesenchymal stem cells).

During the proliferation phase, the primary focus is on the development of granulation tissue, which restores the vascular network and covers the denuded wound surface During this stage, fibroblasts play a substantial and vital role as they multiply in the deeper layers of the wound, synthesising collagen to build a scaffold for migration, fibroblast proliferation, and ECM production. Fibroblasts deposit a lot of extracellular matrix (ECM) under the guidance of cytokines and growth factors such PDGF, FGFs, EGF, TNF, TGF/, CTGF, and IL-1. Furthermore, keratinocytes are induced by a number of cytokines and growth factors, such as EGF, KGFs, FGFs, IL-6, GMCSF, MMPs, and activated protein C, which results in migration, proliferation, stratification, and differentiation. Growth factors like VEGF, PDGF, bFGF, and thrombin promote angiogenesis, which aids in the development of new blood vessels that provide nutrients and oxygen to tissues (10). This phase transitions into the subsequent maturational or remodeling phase, characterized by an immature granulation tissue composed of abundant fibroblasts, M2 macrophages, and loosely organized collagen bundles intertwined with capillaries (9).

During the maturation phase, fibroblast cells migrate away from the wound bed, and collagen undergoes remodeling to form a more organized matrix. The transition from granulation tissue to scar occurs approximately two weeks after the injury. Wound contraction happens due to mechanical tension and the differentiation of fibroblasts into myofibroblasts. Cross-linking of collagen fibrils increases tensile strength, with the assistance of vitamin C. Myofibroblasts, activated by cytokines like TGF-β, express α-smooth muscle actin (SMA) and contract the wound. They also secrete extracellular matrix (ECM), matrix metalloproteinases, and tissue inhibitors of metalloproteinases to remodel the ECM. Collagen III is gradually replaced by collagen I, which has higher tensile strength but takes time to deposit in the ECM. Eventually, myofibroblasts undergo apoptosis once healing is complete. The recovery of skin components such as hair follicles and sweat glands depends on the severity of the wound, and in severe cases, full restoration may not be achieved. Scar tissue, which forms during healing, only reaches 70% to 80% of the original tensile strength after one year following the injury (9, 5,10).

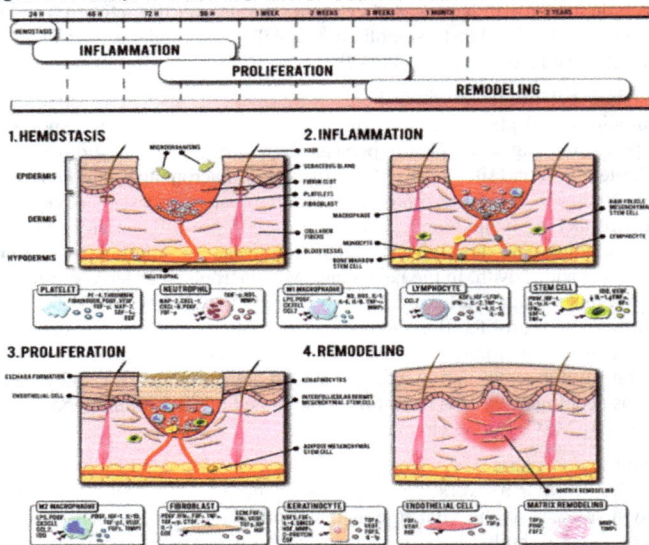

Figure 1 – Different steps in normal wound healing. In cell boxes on left side different mediators like cytokines, GFs (Growth factors) etc. are written that stimulate cells and on right side mediators that are secreted by these cells are written (9).

Chronic/ Delayed wound healing differs significantly from normal wound (acute wound) healing in several aspects. In acute wounds, the healing process typically occurs within a month, and the resulting tissue closely resembles the prewound state. However, chronic wound healing can extend for months or become indefinitely stalled in the inflammation phase. They are commonly linked to underlying conditions like burns, diabetes, vascular disease, or immunological disorders, are marked by impaired circulation, compromised immune response, and persistent inflammation. Notably, infection and inadequate tissue oxygenation further fuel the development of chronic wounds, making their healing process challenging and complex. Insufficient microbial clearance prolongs the inflammatory stage, and the formation of biofilms, which produce an exopolysaccharide matrix (EPS), shields bacteria from the immune response and antibiotics. Additionally, maintaining normoxic conditions is crucial for processes like epithelization, angiogenesis, and collagen deposition during wound healing. It's worth noting that chronic wounds can be identified by the presence of hypoxia, with oxygen levels below 20 mm Hg serving as a reliable indicator. These factors hinder the normal healing cascade and lead to persistent wounds (11).

Like Diabetic wounds display an irregular inflammatory response and decreased neovascularization, as compared to non-diabetic wounds (12). Diabetes, a group of metabolic

Nanomaterials in Biological Systems   Materials Research Forum LLC
Materials Research Foundations 185 (2026) 73-91   https://doi.org/10.21741/9781644903858-4

diseases characterized by blood glucose dysregulation, often leads to the formation of chronic wounds through complex pathophysiological mechanisms. Elevated blood glucose levels, known as hyperglycemia, contribute to increased levels of advanced glycation end-products (AGEs) in the blood. These AGEs promote the generation of reactive oxygen species (ROS) and reactive nitrogen species (RNS), which can cause tissue damage when their levels exceed the tissue's antioxidant capacity. During the inflammation stage of wound healing, ROS and RNS play a crucial role in clearing dead tissue and pathogens. However, excessive concentrations of these reactive species can impede the transition from the inflammatory to the proliferative stage, hindering the formation of new, healthy tissue. Following injury, chronic wounds exhibit a significant rise in proinflammatory cytokine (IL-1 and TNF-α) levels with decreased anti-inflammatory signals (CD206, IGF-1, TGF- and IL-10) will lead to abnormal apoptosis of fibroblasts and keratinocytes, together with decreased angiogenesis. This elevation leads to a prolonged inflammatory stage and excessive production of MMPs (matrix metalloproteinases). The upregulation of MMPs results in the continuous breakdown of extracellular matrix (ECM) components and further degradation of growth factors that are already limited. Consequently, dysregulated cytokine expression poses a significant hindrance to the regrowth of healthy tissue (11).

Similarly burn wounds that are caused by thermal or chemical damage, present complex and challenging healing processes. The severity of burns varies from superficial partial-thickness to full-thickness. Burn wounds commonly experience excessive inflammation and delayed epithelialization. Scar formation is a significant concern, as the loss of skin elasticity and abnormal collagen deposition can lead to hypertrophic scars and contractures. Another distinguishing factor is the presence of inanimate tissues in burn wound beds, which can serve as a medium for potential pathogenic bacteria, increasing the risk of wound infection. Burn wounds exhibit unique characteristics compared to other wounds. Unlike typical wounds, burn wounds do not undergo a homeostasis phase due to the absence of bleeding. Instead, immediately after a burn, increased capillary permeability occurs, leading to fluid leakage from microvessels into the interstitial area. This results in postburn edema and the release of histamine, causing vasoconstriction, decreased intravascular fluid, and hypovolemic shock. The activation of the kinin system further contributes to the inflammation process, affecting the complement cascade, arachidonic acid cascade, and coagulation fibrinolytic cascade. Specifically, increases in arachidonic acid cascade components such as prostaglandins, prostacyclin, and thromboxane A2 perpetuate burn edema formation, vasoconstriction, and local tissue ischemia. This all leads to delayed wound healing (13,14).

This impaired/ delayed wound healing poses a substantial health risk, as it results in wounds that fail to heal within an appropriate timeframe. This not only burdens patients but also has a significant financial impact on healthcare systems worldwide. Globally, the annual cost for wound care escalated from an average of $2.8 billion in 2014 to $3.5 billion in 2021 (15). In the Indian population, the prevalence of acute wounds stands at 4.5 per 1000 individuals, while chronic wounds affect 10.5 per 1000 individuals (16). These statistics shed light on the magnitude of the problem and emphasize the need for effective wound management and prevention strategies.

3.   Conventional treatments

Traditional wound dressings were primarily used as passive barriers to protect wounds from external contamination. However, advancements in wound dressing technology have led to the development of more sophisticated dressings capable of creating a protective environment and delivering active compounds. Currently, there is a wide range of wound dressing materials available for clinical use, including hydrogels, foam, film, antibacterial creams, ointments, and combinations of antibacterial agents with polymers. An ideal wound dressing should mimic the extracellular matrix, offering biological stability, flexibility, and moisture management while protecting against external dangers, bacterial infections, and promoting tissue healing. It should also regulate temperature, facilitate epidermal migration, angiogenesis, and connective tissue synthesis, while being hypoallergenic and cost-effective. While systemically administered antibiotics may have limited penetration into wound biofilms, their topical application has proven effective. Commonly used antiseptics such as iodine, chlorhexidine, ethacridine, and silver, along with antibiotics like gentamicin, tetracycline, chloramphenicol, vancomycin, framycetin, and mupirocin, offer clear benefits in wound care. Additionally, growth factors (EGF, PDGF, bFGF, GM-CSF, TGF-β), which are biologically active peptides, regulate cell functions and are being studied for topical treatments in various forms (gels, creams, injections). These growth factors demonstrate notable effects on skin healing with minimal side effects. For sterile wounds, dressings that maintain a moist environment and allow for rapid re-epithelialization, such as hydrocolloids, semi-permeable polyurethane, and calcium alginate dressings, are advised. However, the treatment of necrotic or infected wounds necessitates the use of "functionalized" dressings containing antibiotics or other antibacterial chemicals (17,18,9,19,5).However these conventional methods of wound healing have certain drawbacks that can impede the healing process. For example these commonly available wound dressings don't maintain a moist environment, they have poor absorption of wound exudates and cause poor gas exchange between wound and the environment that leads to delayed wound healing and difficulty in removal of the wound dressing after healing. It is also crucial to highlight that repeated and incorrect antibiotic usage can result in the selection of antibiotic-resistant bacteria, and roughly 70% of bacteria in chronic wounds are resistant to at least one commonly administered antibiotic. Ultimately, conventional methods often prove inadequate in effectively addressing chronic wounds or wounds associated with underlying medical conditions, leading to prolonged healing periods and an increased risk of complications.

There is an urgent need for alternative therapies that can offer precise outcomes in the wound-healing process, including wound closure and fluid control, while possessing properties such as durability, elasticity, and biocompatibility. Nanomaterial-based approaches have emerged as successful strategies in combating bacterial infections and providing additional benefits, such as cell-type specificity, which conventional wound-dressing materials lacked. Nanomaterials have found applications in various biomedical fields due to their nanoscale size, large surface area-to-volume ratio, and favourable physicochemical properties. Their ability to efficiently interact with wound areas and penetrate the skin layer makes them valuable not only as therapeutic agents for wound healing but also for sustained and controlled release of therapeutics. Thus, nanomaterials hold significant potential for enhancing wound healing and finding utility in diverse biomedical applications (20). Current wound healing therapies generally fail to provide a good clinical outcome, either structurally (e.g., wound re-epithelialization, control of fluid loss), or functionally (e.g., histological features that determine elasticity, durability, sensitivity, etc.). For this reason, nanotechnology, through the versatile physicochemical properties, is a reliable research domain for wound-healing therapies (19). In order to accelerate wound closure and restore functionality to healed tissue, advanced dressings should utilize biodegradable and biocompatible materials that target specific events in the healing process. Nanoparticles (NPs), as part of nanomedicine tools, offer promising strategies to enhance tissue regeneration therapies and drive innovation in this field.

Table 1 – Different conventional treatment for wound healing along with their advantages and disadvantages

Conventional treatments	Advantages	Disadvantages	References
Woven gauze	Absorbs wound exudate Allows the wound to form an eschar	Difficult to remove Cause mechanical debridement and/or wound trauma Formation of granuloma due to woven gauze residuals	(21, 22).
Hydrogels	Available as an amorphous gel in a tube (used in cavity wound), Flexible sheets, and hydrogel-impregnated gauze ↑ autolytic debridement via the rehydration of eschar and slough Nonadherent Relieve pain Create and maintain a moist environment	Maceration occurs in case of high-exudate wounds Amorphous gels require a secondary dressing Need frequent application	(21, 23, 24).
Traditional/passive dressings (e.g., gauze, bandages, wool dressings, and plaster)	Protect the wound from foreign impurities and other contaminants Control bleeding Absorb exudate Provide cushion	Require frequent changing Do not offer a moist wound environment Damage the skin and cause pain as well as bacterial invasion	(25, 26).
Silver dressing (e.g., foams, alginates, films, mesh, hydrocolloids and gels)	Broad-spectrum antibacterial	May require a secondary dressing Transient skin staining Burning sensation Contraindicated in patients who are sensitive to silver	(23).
Sponges	High porosity Thermal insulation Sustain a moist environment Absorb wound exudates Enhance tissue regeneration	Mechanically weak May provoke skin maceration Unsuitable for third-degree burn treatment or wounds with dry eschar	(27, 28).
Gentamicin	Cream, 0.1%; and ointment, 0.1%, Broad spectrum Inexpensive	Must be applied 3-4 times daily; May drive resistance to an agent used systemically	(29).
Povidone iodine	Ointment, 1%, 4.7%, 10%; solution Broad spectrum Less irritating to skin and allergenic than iodine. Can be covered with dressings. Clinically significant resistance very rare	Antibacterial action requires at least 2 min contact; May cause stinging and erythema; Effect may not persist, and efficacy may be reduced in body fluids;	(29).

4. Nanotechnology

Nanotechnology focuses on the synthesis, structure, and dynamics of atomic and molecular particles at the nanoscale, specifically nanoparticles with a maximum diameter of 100 nm. The term "nanoparticle" (NP) is defined by the International Organization for Standardization as a particle with a size ranging from 1 to 100 nm. However, in common usage, "nanoparticle" is often used to describe any particle smaller than 1 mm (30).Various analytical techniques, such as dynamic light scattering (DLS), fluorescence correlation spectroscopy (FCS), Raman scattering (RS), circular dichroism (CD), infrared spectroscopy (IR), mass spectroscopy (MS), transmission electron microscopy (TEM), and scanning electron microscopy (SEM), are employed to investigate the size, structure, conformation, surface characteristics, and morphology of nanomaterials. These properties, including size and shape, play a crucial role in determining the biological efficiency of nanoparticles, affecting factors such as drug delivery, penetrability, and cellular responses (19). In the medical field, nanotechnology has attracted considerable interest in drug delivery, diagnosis, and therapeutic applications due to the inherent ability of nano-delivery systems to (I) protect the drug from the environment by physical and chemical means, (II) control and sustained release of cargo, (III) improve the solubility of hydrophobic drugs, (IV) enable a targeted delivery to specific cells and tissues by functionalizing their surfaces with cell-specific ligands, and (V) preferentially accumulate into target tissues due to the enhanced permeability and retention (EPR) effect (31).

The larger surface area to volume ratio of nanomaterials enables improved cell adhesion and the potential to encapsulate a higher number of surface-functionalized active components, thereby enhancing specific regenerative functions. Utilizing nanotechnology in wound healing provides advantages such as localized drug delivery, cell specificity, and sustained and controlled release of encapsulated drugs for the necessary duration until the wound heals. Specifically, nanoparticles possess optimal characteristics for topical administration, promoting enhanced interactions with biological targets and improved penetration at the wound sites (20).The inherent bactericidal and fungicidal properties of certain nanoparticles (NPs) have led to their utilization in various therapeutic approaches, showcasing effectiveness in wound care and related biomedical applications. This also highlights the significance of nanomedicine as a valuable avenue for the development of novel antimicrobial agents (17).

5. How nanoparticles will help in wound healing?

The unique properties of nanomaterials (NMs) introduce a new avenue for the development of wound-healing products. NMs possess antibacterial, anti-inflammatory, proangiogenic, and proliferative properties, enabling them to influence every stage of the wound healing process. Additionally, NMs have the ability to modulate the expression of vital proteins and signaling molecules, thereby enhancing the wound healing process. Consequently, NMs, either alone or in combination with materials at both micro- and nanoscales, hold great promise for overcoming the challenges encountered in wound care management. The primary categories of nanomaterials suitable for wound treatment include nanoparticles, nanocapsules, nanocomposites, coatings, and scaffolds (32). In wound therapy, two primary categories of nanoparticles (NPs) are frequently utilized: (1) NPs with inherent properties that facilitate wound closure, and (2) NPs utilized as carriers for therapeutic substances. The first category can be further classified into metallic/metal

Nanomaterials in Biological Systems   Materials Research Forum LLC
Materials Research Foundations 185 (2026) 73-91   https://doi.org/10.21741/9781644903858-4

oxide nanomaterials and nonmetallic nanomaterials (19). Some type of nanomaterial that are used in wound healing are -

Nanospheres are miniature polymeric matrix systems that consist of a solid porous polymer capable of binding active substances such as amino acids, minerals, or organic compounds. This binding process enhances the stability of the active compounds and provides them with enhanced biocompatibility and improved pharmaceutical properties. Chitosan, polylactic acid (PLA), gelatin, and polylactide/glycolide are commonly employed polymers in the composition of nanospheres and active compound is released through diffusion, particularly for water-soluble agents. Muller designed three-dimensional (3D) electrospun poly(D,L-lactide) (PLA) fibermats. These fibermats were enhanced with the addition of nanospheres, skillfully formed from amorphous calcium polyphosphate (polyP) nanoparticles (NP) that contained encapsulated retinol, delightfully termed 'retinol/aCa-polyP-NS' nanospheres (NS) that help in acceerated wound healing (33).

Nanocapsules possess the ability to encapsulate an active agent within their structure and release it at predetermined intervals, thereby ensuring a more precise and controlled delivery. They also enhance the penetration of active ingredients into the deeper layers of the dermis. However, it is important to note that Qthis may potentially lead to a decrease in the effectiveness of the substance. Quartinello designed an antibacterial wound dressing using pH-responsive human serum albumin/silk fibroin nanocapsules immobilized on cotton/polyethylene terephthalate (PET) blends loaded with eugenol. This innovative dressing demonstrated strong antimicrobial properties against Staphylococcus aureus and Escherichia coli strains, effectively preventing bacterial migration in wounds and accelerating the healing process (34).

Nanoemulsions are uniform and thermally stable emulsions with an oil-in-water composition, featuring droplets that are no larger than 100 nm in size. These emulsions include surfactants and have the ability to effectively incorporate biologically active substances, preventing them from sedimenting. Nanoemulsions are known for their high solubilization capacity, fluidity, and reduced viscosity. However, a potential drawback is the unregulated accumulation of active substances in the reticular dermis or subcutaneous fat. Parham and Kharazi developed nanoemulsion of clove oil, revealing remarkable wound healing effects in rats when compared to pure clove oil. Additionally, the nanoemulsion exhibited potent antimicrobial activity against various bacterial strains, including E. coli, P. aeruginosa, S. epidermidis, and S. aureus (35).

Nanocarriers exhibit a similar structure to liposomes, but are composed of synthetic, non-ionic surfactants. Moreover, they possess surface receptors that have the ability to bind to specific target sites. This characteristic enhances the effectiveness of the active agent while minimizing the occurrence of adverse reactions.

Nanocolloids consist of non-ionic metal nanoparticles ranging from 1.5 to 5 nm in size, dispersed extensively in demineralized water. The perpetual Brownian motion of these nanoparticles enables them to easily penetrate both eukaryotic and prokaryotic cells, thereby exhibiting antimicrobial properties that is good for early wound healing. Additionally, nanocolloids demonstrate favorable electrical conductivity and are positively charged particles.

Scaffold refers to a three-dimensional structure or matrix, often made of biomaterials, designed to support and facilitate the regeneration of tissue at the wound site. The scaffold serves as a framework that mimics the extracellular matrix (ECM) of the tissue, providing a suitable environment for cells to migrate, proliferate, and differentiate during the healing process. Losi discovered that their scaffold, which consisted of poly(ether)urethane-polydimethylsiloxane/fibrin and incorporated PLGA nanoparticles loaded with VEGF and basic fibroblast growth factor (bFGF) (scaffold/growth factor-loaded NPs), effectively stimulated granulation tissue formation, collagen secretion, and re-epithelialization. As a result, the wound closure rate significantly increased in diabetic mice when compared to scaffolds containing PLGA nanoparticles without growth factors and the control group (36).

Solid lipid nanoparticles (SLNs) and nanostructured lipid carriers (NLCs) are representatives of lipid nanoparticles. SLNs are colloidal carriers having size range between 50 and 1000 nm with a lipid-forming core at both body and room temperatures. SLNs could be modified by incorporating lipid liquid into the solid structure forming NLCs, conferring better loading capacity and more stable composition. Gainza utilized the emulsion ultrasonication method to create solid lipid nanoparticles (SLN) and nanostructured lipid carriers (NLC) loaded with recombinant human epidermal growth factor (rhEGF). Their study demonstrated that SLN-rhEGF and NLC-rhEGF led to significantly improved wound closure in diabetic mice compared to free rhEGF and alginate microspheres containing rhEGF. This suggests that the lipid nanoparticles enable controlled release of rhEGF while preserving its bioactivity upon encapsulation, highlighting their potential for effective wound healing applications (37).

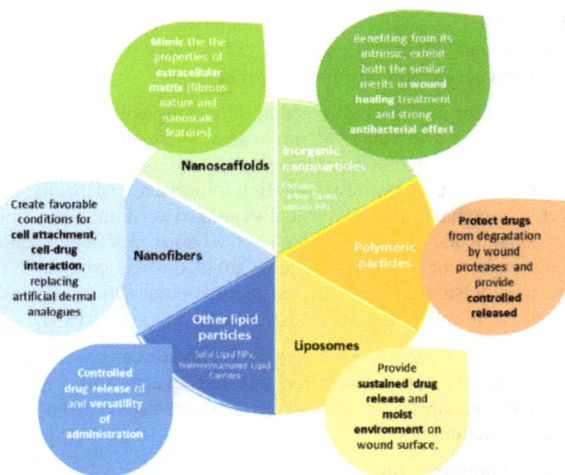

Figure 2 – Key properties of several nano-based systems utilised in wound healing (54).

Nanomaterials offer unique properties that make them highly suitable for applications in wound care. These materials can be engineered to provide localized and controlled drug delivery, promoting tissue regeneration and reducing infection risk. Their small size allows for improved interactions with biological targets, facilitating enhanced wound healing processes. As a result, incorporating nanomaterials into wound dressings and therapeutic agents has the potential to significantly enhance wound healing outcomes, leading to better patient recovery and overall quality of life. The key benefits of employing NPs include the ability to deliver medicines, genes, and peptides at elevated amounts with very few adverse effects as compared to conventional drug delivery methods, increasing the efficacy of the therapy

For example - Metallic nanoparticles (MNPs) produced from inorganic sources proved to be more effective when compared to organic nanomaterials due to their unique intrinsic physicochemical properties. Their small size and high reactivity enable them to penetrate deeper into the wound site and afford better association with biological components. Due to their high surface-area-to-volume ratio, they can serve as a vehicle for other therapeutics, leading to accelerated and synergistic healing properties. Moreover, MNPs have enhanced mechanical strength, can facilitate controlled drug release and have been shown to have toxic effects towards both fungi and bacteria, these properties further confirm them as excellent candidates for various medical applications, including wound healing. These metallic NPs is capable of developing a new therapeutic modality in the treatment of wounds, demonstrating potent effects in reducing infections caused by microorganisms and in reducing healing time, reducing the damage caused by the inflammatory process chronic. The binding of metallic NPs with biological molecules, such as hyaluronic acid (HA), optimizes the secretion of anti-inflammatory cytokines, proliferation, and growth factors of cell differentiation and makes an earlier transition to the chronic phase, contributing to the process of repair (38). Among these, AgNPs are the most researched, highly utilised and commercialised nanomaterials for various biomedical purposes, especially for applications in wound care (39). Due to their antibacterial properties and low toxicity profile,metal NPs such as copper, silver, gold and zinc represent ideal candidates for integration in wound dressings (40).

Polymeric nanoparticles, such as chitosan, cellulose or PLGA loaded nanoparticles, are utilized in polymeric nanomaterial therapy either as wound dressings or delivery carriers. These nanoparticles possess antibacterial properties and promote re-epithelialization, making them suitable for wound treatment. Biopolymers, which are biocompatible polymeric networks, can absorb large amounts of fluid while maintaining a moist wound environment (19). Hajji demonstrated that chitosan-polyvinyl alcohol-silver nanoparticles exhibited enhanced antioxidant and antimicrobial properties compared to the simple chitosan polymer. The nanocomplex showed superior effectiveness against both Gram-positive (Staphylococcus aureus) and Gram-negative (Escherichia coli and Klebsiella pneumoniae) bacteria. Moreover, in vivo studies revealed that this nanocomplex significantly improved wound closure by promoting granulation and re-epithelialization while displaying reduced cytotoxicity (41). Holban found that coatings comprising polylactic acid-chitosan-magnetite-eugenol nanospheres effectively suppressed biofilm synthesis of Staphylococcus aureus and Pseudomonas aeruginosa. Additionally, these coatings promoted endothelial proliferation, demonstrating their potential in combating bacterial infections and supporting wound healing (42). Sun used carboxymethyl chitosan nanoparticles to encapsulate cathelicidin, an antimicrobial peptide extracted from king cobra, and studied its impact on wound healing. Cathelicidin exhibited strong antibacterial properties, especially in the presence of 1% NaCl. Additionally, it regulated cytokine release, promoting anti-inflammatory IL-10 expression and reducing pro-inflammatory cytokines like TNF-α, IL-1β, and IL-6. Interestingly, the cathelicidin-carboxymethyl chitosan nanoparticles enhanced cell migration without affecting keratinocyte growth. Moreover, during the later stages of wound closure, the collagen I to III ratio favored non-scarring healing (43).

Nanomaterials in Biological Systems   Materials Research Forum LLC
Materials Research Foundations 185 (2026) 73-91   https://doi.org/10.21741/9781644903858-4

Moniri created nanocomposites by combining bacterial nanocellulose derived from the fermentation of Gram-negative Gluconacetobacter xylinus with silver nanoparticles (BNC-Ag). These nanocomposites demonstrated enhanced wound healing properties and showed reduced Staphylococcus aureus colonization in vitro (44).

Table 2 – Different types of nanomaterials are involved in wound healing mechanisms, and their effects on wound healing are discussed.

Type of Nanoparticle	Material	Wound healing mechanism	References
Metal and metal oxide nanoparticle	Silver nanoparticles	Have antibacterial activity, Antiinflammatory activities Increase reepithelialisation Increase wound closure Increase ROS Increase bacterial clearance from infected wounds Decrease cytokine release Decrease lymphocytic infiltration Decrease mast cell infiltration	(21,38,45)
	Gold nanoparticles	High chemical stability Simple synthesis Absorb near infrared (NIR) light ↓ activity of multidrug resistant (MDR) bacteria ↓ bacterial adenosine triphosphate (ATP) Antioxidant ↑ transcutaneous codelivery of other drugs ↑ collagen deposition ↑ reepithelialisation ↑ granulation tissue formation ↑ wound closure	(21,38,17,46).
	Copper nanoparticles	↑ wound closure Regulate cells, cytokines, and growth factor involved in wound healing ↑ proliferation of keratinocytes and fibroblasts ↑ epithelialisation ↑ collagen synthesis ↑ angiogenesis ↑ transforming growth factor beta (TGF-β), VEGF, matrix metalloproteinase-2 (MMP-2), hypoxia-inducible factor 1-alpha (HIF-1α), interleukin-10 (IL-10), and tumour necrosis factor alpha (TNF-α) Extracellular matrix remodelling	(38,47,48).
	Zinc oxide nanoparticles	Bactericidal and bacteriostatic ↑ ROS and apoptosis ↑ aggregate stability ↓ toxicity	(38,18,20,49).

		↑ permeability via the skin and lipid membranes ↑ collagen synthesis	
Polymeric nanoparticles	Curcumin-PLGA	Anti-inflammatory activity, improved reepithelialization and granulation tissue formation	(50,51).
	PLGA NPs loaded with VEGF, bFGF	VEGF and bFGF loaded nanoparticles treated wound stimulated significant granulation tissue formation, collagen secretion and re-epithelialization, and accelerated wound closure compared to controls and NPs without biomolecules	(52,36).
	Enoxaparin-PCL-chitosan	Improved wound healing via correction of hemostasis and inflammation	(50).
	Curcumin-TMSO-PEG chitosan TMSO: Tetramethylorthosilicate	Antimicrobial effect Accelerated wound closure via well-formed granulation tissue Enhanced collagen deposition New vessel formation and reepithelialization	(50).
	NO-TMSO-PEG chitosan	Stimulated migration and proliferation of fibroblasts, and collagen type III expression, increase vascularization, less inflammation, upregulated expression of genes associated with extracellular matrix formation and VEGF	(53).
	Quercetin loaded chitosan tripolyphosphate nanoparticles	Better wound healing in rats by controlled modulations of cytokines (TNF-α and IL-10) and growth factors (VEGF and TGF-β1) involved in inflammatory and proliferative phases of cutaneous wound healing.	(54).
	Bilirubin-encapsulated pluronic F-127 nanoparticles	Improved the quality of cutaneous wound healing in rats by controlled modulations of different cytokines (TNF-α and IL-10) and growth factors (VEGF and TGF-β1) involved in inflammatory and proliferative phases of wound healing.	(55).
Solid Lipid Nanoparticles	Morphine-loaded solid lipid nanoparticles	High encapsulation efficiency and high drug loading and sustained release; Promoted re-epithelialization; Enhanced proliferation and migration of keratinocytes and the drug uptake by fibroblasts; Angiogenetic effects on wound	(56,57).

		repair; Inhibitory efficacy on scar complications.	
	rhEGF-loaded lipid nanoparticles	Effectiveness in enhanced wound closure after topical administration in diabetic mice; Formation of new connective tissue and blood vessels.	(37).
Antibiotic-loaded nanoparticles (nanobiotics)	Vancomycin-modified nanoparticles	Broad spectrum antibacterial activity ↑ actions of antibiotics against MDR microorganisms ↑ tissue regeneration ↓ toxicity	(37).
	Epigallocatechin gallate nanoparticles	Comprised activated carbon fibers with gentamicin that enhance wound healing	(58).
Scaffold	Scaffold loaded with Silver nanoparticles	Strong anti-microbial capability and excellent biocompatibility with fibroblast cells	(59,60).
	PDGF-BB and VEGF	Accelerated tissue regeneration and remodeling, promoted angiogenesis	(59,61).

6. Future prospects and conclusion

Nanoparticle-based approaches for wound healing hold immense promise and present exciting future prospects in the field of biomedical research. With ongoing advancements in nanotechnology, researchers are continually exploring novel strategies to utilize nanoparticles for enhanced wound healing outcomes. These approaches offer the potential for targeted and controlled delivery of therapeutic agents, such as growth factors, antimicrobial agents, and anti-inflammatory drugs, directly to the wound site. Moreover, nanoparticles can be engineered with specific physicochemical properties to optimize interactions with biological tissues and promote cellular responses crucial for tissue regeneration. The development of smart and multifunctional nanoparticles holds the key to addressing various challenges in wound healing, such as infection control, accelerated healing, scar reduction, and promoting tissue regeneration. Additionally, the integration of nanomaterials with emerging technologies, such as 3D printing and tissue engineering, opens up new avenues for the creation of advanced wound dressings and scaffolds that mimic the native extracellular matrix. As researchers continue to delve deeper into the potential of nanoparticle-based approaches, we can anticipate transformative breakthroughs that will revolutionize wound healing therapies and improve patient outcomes in the future.

References

[1] K. Vig, A. Chaudhari, S. Tripathi, S. Dixit, R. Sahu, S. Pillai, V.A. Dennis, S.R. Singh, Advances in skin regeneration using tissue engineering, Int. J. Mol. Sci. 18 (2017). https://doi.org/10.3390/ijms18040789

[2] R. Wong, S. Geyer, W. Weninger, J.C. Guimberteau, J.K. Wong, The dynamic anatomy and patterning of skin, Exp. Dermatol. 25 (2016) 92–98. https://doi.org/10.1111/exd.12832

[3]A. Nourian Dehkordi, F. Mirahmadi Babaheydari, M. Chehelgerdi, S. Raeisi Dehkordi, Skin tissue engineering: Wound healing based on stem-cell-based therapeutic strategies, Stem Cell Res. Ther. 10 (2019). https://doi.org/10.1186/s13287-019-1212-2

[4]S. Saghazadeh, C. Rinoldi, M. Schot, S.S. Kashaf, F. Sharifi, E. Jalilian, K. Nuutila, G. Giatsidis, P. Mostafalu, H. Derakhshandeh, K. Yue, W. Swieszkowski, A. Memic, A. Tamayol, A. Khademhosseini, Drug delivery systems and materials for wound healing applications, Adv. Drug Deliv. Rev. 127 (2018) 138–166. https://doi.org/10.1016/j.addr.2018.04.008

[5]I. Negut, G. Dorcioman, V. Grumezescu, Scaffolds for wound healing applications, Polymers (Basel). 12 (2020) 1–19. https://doi.org/10.3390/polym12092010

[6]E. Rezvani Ghomi, S. Khalili, S. Nouri Khorasani, R. Esmaeely Neisiany, S. Ramakrishna, Wound dressings: Current advances and future directions, J. Appl. Polym. Sci. 136 (2019). https://doi.org/10.1002/app.47738

[7]A. Moeini, P. Pedram, P. Makvandi, M. Malinconico, G. Gomez d'Ayala, Wound healing and antimicrobial effect of active secondary metabolites in chitosan-based wound dressings: A review, Carbohydr. Polym. 233 (2020). https://doi.org/10.1016/j.carbpol.2020.115839

[8]M. Rodrigues, N. Kosaric, C.A. Bonham, G.C. Gurtner, Wound Healing: A Cellular Perspective, Physiol Rev. 99 (2019) 665–706. https://doi.org/10.1152/physrev.00067.2017.-Wound

[9]K. Las Heras, M. Igartua, E. Santos-Vizcaino, R.M. Hernandez, Chronic wounds: Current status, available strategies and emerging therapeutic solutions, J. Control. Release. 328 (2020) 532–550. https://doi.org/10.1016/j.jconrel.2020.09.039

[10]   S. Muzammil, J. Neves Cruz, R. Mumtaz, I. Rasul, S. Hayat, M.A. Khan, A.M. Khan, M.U. Ijaz, R.R. Lima, M. Zubair, Effects of Drying Temperature and Solvents on In Vitro Diabetic Wound Healing Potential of Moringa oleifera Leaf Extracts, Molecules. 28 (2023). https://doi.org/10.3390/molecules28020710

[11]   M.J. Malone-Povolny, S.E. Maloney, M.H. Schoenfisch, Nitric Oxide Therapy for Diabetic Wound Healing, Adv. Healthc. Mater. 8 (2019) 1–18. https://doi.org/10.1002/adhm.201801210

[12]   D.S. Masson-Meyers, T.A.M. Andrade, G.F. Caetano, F.R. Guimaraes, M.N. Leite, S.N. Leite, M.A.C. Frade, Experimental models and methods for cutaneous wound healing assessment, Int. J. Exp. Pathol. 101 (2020) 21–37. https://doi.org/10.1111/iep.12346

[13]   Y. Yao, A. Zhang, C. Yuan, X. Chen, Y. Liu, Recent trends on burn wound care: Hydrogel dressings and scaffolds, Biomater. Sci. 9 (2021) 4523–4540. https://doi.org/10.1039/d1bm00411e

[14]   L.G. Wasef, H.M. Shaheen, Y.S. El-Sayed, T.I.A. Shalaby, D.H. Samak, M.E. Abd El-Hack, A. Al-Owaimer, I.M. Saadeldin, A. El-mleeh, H. Ba-Awadh, A.A. Swelum, Effects of Silver Nanoparticles on Burn Wound Healing in a Mouse Model, Biol. Trace Elem. Res. 193 (2020) 456–465. https://doi.org/10.1007/s12011-019-01729-z

[15]   C.K. Sen, Human Wounds and Its Burden: An Updated Compendium of Estimates, Adv. Wound Care. 8 (2019) 39–48. https://doi.org/10.1089/wound.2019.0946

[16] V.K. Shukla, M.A. Ansari, S.K. Gupta, Wound healing research: A perspective from India, Int. J. Low. Extrem. Wounds. 4 (2005) 7–8. https://doi.org/10.1177/1534734604273660

[17] M.H. Sarfraz, M. Zubair, B. Aslam, A. Ashraf, M.H. Siddique, S. Hayat, J.N. Cruz, S. Muzammil, M. Khurshid, M.F. Sarfraz, A. Hashem, T.M. Dawoud, G.D. Avila-Quezada, E.F. Abd_Allah, Comparative analysis of phyto-fabricated chitosan, copper oxide, and chitosan-based CuO nanoparticles: antibacterial potential against Acinetobacter baumannii isolates and anticancer activity against HepG2 cell lines, Front. Microbiol. 14 (2023). https://doi.org/10.3389/fmicb.2023.1188743

[18] N.K. Rajendran, S.S.D. Kumar, N.N. Houreld, H. Abrahamse, A review on nanoparticle based treatment for wound healing, J. Drug Deliv. Sci. Technol. 44 (2018) 421–430. https://doi.org/10.1016/j.jddst.2018.01.009

[19] M.M. Mihai, M.B. Dima, B. Dima, A.M. Holban, Nanomaterials for wound healing and infection control, Materials (Basel). 12 (2019). https://doi.org/10.3390/ma12132176

[20] J.N. Cruz, S. Muzammil, A. Ashraf, M.U. Ijaz, M.H. Siddique, R. Abbas, M. Sadia, Saba, S. Hayat, R.R. Lima, A review on mycogenic metallic nanoparticles and their potential role as antioxidant, antibiofilm and quorum quenching agents, Heliyon. 10 (2024). https://doi.org/10.1016/j.heliyon.2024.e29500

[21] M.A. Shalaby, M.M. Anwar, H. Saeed, Nanomaterials for application in wound Healing: current state-of-the-art and future perspectives, J. Polym. Res. 29 (2022). https://doi.org/10.1007/s10965-021-02870-x

[22] A.J. Rosenbaum, S. Banerjee, K.M. Rezak, R.L. Uhl, Advances in wound management, J. Am. Acad. Orthop. Surg. 26 (2018) 833–843. https://doi.org/10.5435/JAAOS-D-17-00024

[23] D. Bell, D. Hyam, Choosing an appropriate dressing for chronic wounds, Prescriber. 18 (2007) 65–70. https://doi.org/10.1002/psb.89

[24] T.K. Dash, V.B. Konkimalla, Poly-ε-caprolactone based formulations for drug delivery and tissue engineering: A review, J. Control. Release. 158 (2012) 15–33. https://doi.org/10.1016/j.jconrel.2011.09.064

[25] S. Dhivya, V.V. Padma, E. Santhini, Wound dressings - A review, Biomed. 5 (2015) 24–28. https://doi.org/10.7603/s40681-015-0022-9

[26] B.S. Anisha, R. Biswas, K.P. Chennazhi, R. Jayakumar, Chitosan-hyaluronic acid/nano silver composite sponges for drug resistant bacteria infected diabetic wounds, Int. J. Biol. Macromol. 62 (2013) 310–320. https://doi.org/10.1016/j.ijbiomac.2013.09.011

[27] M.A. Matica, F.L. Aachmann, A. Tøndervik, H. Sletta, V. Ostafe, Chitosan as a wound dressing starting material: Antimicrobial properties and mode of action, Int. J. Mol. Sci. 20 (2019). https://doi.org/10.3390/ijms20235889

[28] R. Jayakumar, M. Prabaharan, P.T. Sudheesh Kumar, S. V. Nair, H. Tamura, Biomaterials based on chitin and chitosan in wound dressing applications, Biotechnol. Adv. 29 (2011) 322–337. https://doi.org/10.1016/j.biotechadv.2011.01.005

[29]   B.A. Lipsky, C. Hoey, M. Cruciani, C. Mengoli, Topical antimicrobial agents for preventing and treating foot infections in people with diabetes, Cochrane Database Syst. Rev. 2014 (2014). https://doi.org/10.1002/14651858.CD011038

[30]   M. Berthet, Y. Gauthier, C. Lacroix, B. Verrier, C. Monge, Nanoparticle-Based Dressing: The Future of Wound Treatment?, Trends Biotechnol. 35 (2017) 770–784. https://doi.org/10.1016/j.tibtech.2017.05.005

[31]   E.A. Madawi, A.R. Al Jayoush, M. Rawas-Qalaji, H.E. Thu, S. Khan, M. Sohail, A. Mahmood, Z. Hussain, Polymeric Nanoparticles as Tunable Nanocarriers for Targeted Delivery of Drugs to Skin Tissues for Treatment of Topical Skin Diseases, Pharmaceutics. 15 (2023). https://doi.org/10.3390/pharmaceutics15020657

[32]   A.E. Stoica, C. Chircov, A.M. Grumezescu, Nanomaterials for wound dressings: An Up-to-Date overview, Molecules. 25 (2020). https://doi.org/10.3390/molecules25112699

[33]   W.E.G. Müller, E. Tolba, B. Dorweiler, H.C. Schröder, B. Diehl-Seifert, X. Wang, Electrospun bioactive mats enriched with Ca-polyphosphate/retinol nanospheres as potential wound dressing, Biochem. Biophys. Reports. 3 (2015) 150–160. https://doi.org/10.1016/j.bbrep.2015.08.007

[34]   F. Quartinello, C. Tallian, J. Auer, H. Schön, R. Vielnascher, S. Weinberger, K. Wieland, A.M. Weihs, A. Herrero-Rollett, B. Lendl, A.H. Teuschl, A. Pellis, G.M. Guebitz, Smart textiles in wound care: Functionalization of cotton/PET blends with antimicrobial nanocapsules, J. Mater. Chem. B. 7 (2019) 6592–6603. https://doi.org/10.1039/c9tb01474h

[35]   S. Parham, A.Z. Kharazi, H.R. Bakhsheshi-Rad, H. Nur, A.F. Ismail, S. Sharif, S. Ramakrishna, F. Berto, Antioxidant, antimicrobial and antiviral properties of herbal materials, Antioxidants. 9 (2020) 1–36. https://doi.org/10.3390/antiox9121309

[36]   P. Losi, E. Briganti, C. Errico, A. Lisella, E. Sanguinetti, F. Chiellini, G. Soldani, Fibrin-based scaffold incorporating VEGF- and bFGF-loaded nanoparticles stimulates wound healing in diabetic mice, Acta Biomater. 9 (2013) 7814–7821. https://doi.org/10.1016/j.actbio.2013.04.019

[37]   G. Gainza, M. Pastor, J.J. Aguirre, S. Villullas, J.L. Pedraz, R.M. Hernandez, M. Igartua, A novel strategy for the treatment of chronic wounds based on the topical administration of rhEGF-loaded lipid nanoparticles: In vitro bioactivity and in vivo effectiveness in healing-impaired db/db mice, J. Control. Release. 185 (2014) 51–61. https://doi.org/10.1016/j.jconrel.2014.04.032

[38]   C. Mendes, A. Thirupathi, M.E.A.B. Corrêa, Y. Gu, P.C.L. Silveira, The Use of Metallic Nanoparticles in Wound Healing: New Perspectives, Int. J. Mol. Sci. 23 (2022). https://doi.org/10.3390/ijms232315376

[39]   Z.B. Nqakala, N.R.S. Sibuyi, A.O. Fadaka, M. Meyer, M.O. Onani, A.M. Madiehe, Advances in nanotechnology towards development of silver nanoparticle-based wound-healing agents, Int. J. Mol. Sci. 22 (2021). https://doi.org/10.3390/ijms222011272

[40]   J. Salvo, C. Sandoval, Role of copper nanoparticles in wound healing for chronic wounds: Literature review, Burn. Trauma. 10 (2022). https://doi.org/10.1093/burnst/tkab047

[41]   F.S. Alves, J.N. Cruz, I.N. de Farias Ramos, D.L. do Nascimento Brandão, R.N. Queiroz, G.V. da Silva, G.V. da Silva, M.F. Dolabela, M.L. da Costa, A.S. Khayat, J. de Arimatéia Rodrigues do Rego, D. do Socorro Barros Brasil, Evaluation of Antimicrobial Activity and Cytotoxicity Effects of Extracts of Piper nigrum L. and Piperine, Separations. 10 (2023). https://doi.org/10.3390/separations10010021

[42]   A.M. Holban, V. Grumezescu, A.M. Grumezescu, B.Ş. Vasile, R. Truşcă, R. Cristescu, G. Socol, F. Iordache, Antimicrobial nanospheres thin coatings prepared by advanced pulsed laser technique, Beilstein J. Nanotechnol. 5 (2014) 872–880. https://doi.org/10.3762/bjnano.5.99

[43]   T. Sun, B. Zhan, W. Zhang, D. Qin, G. Xia, H. Zhang, M. Peng, S.A. Li, Y. Zhang, Y. Gao, W.H. Lee, Carboxymethyl chitosan nanoparticles loaded with bioactive peptide OH-CATH30 benefit nonscar wound healing, Int. J. Nanomedicine. 13 (2018) 5771–5786. https://doi.org/10.2147/IJN.S156206

[44]   M. Moniri, A.B. Moghaddam, S. Azizi, R.A. Rahim, S.W. Zuhainis, M. Navaderi, R. Mohamad, In vitro molecular study of wound healing using biosynthesized bacteria nanocellulose/ silver nanocomposite assisted by bioinformatics databases, Int. J. Nanomedicine. 13 (2018) 5097–5112. https://doi.org/10.2147/IJN.S164573

[45]   A.C. Burduşel, O. Gherasim, A.M. Grumezescu, L. Mogoantă, A. Ficai, E. Andronescu, Biomedical applications of silver nanoparticles: An up-to-date overview, Nanomaterials. 8 (2018) 1–25. https://doi.org/10.3390/nano8090681

[46]   V. Vijayakumar, S.K. Samal, S. Mohanty, S.K. Nayak, Recent advancements in biopolymer and metal nanoparticle-based materials in diabetic wound healing management, Int. J. Biol. Macromol. 122 (2019) 137–148. https://doi.org/10.1016/j.ijbiomac.2018.10.120

[47]   M. Tiwari, K. Narayanan, M.B. Thakar, H. V. Jagani, J.V. Rao, Biosynthesis and wound healing activity of copper nanoparticles, IET Nanobiotechnology. 8 (2014) 230–237. https://doi.org/10.1049/iet-nbt.2013.0052

[48]   A. Gopal, V. Kant, A. Gopalakrishnan, S.K. Tandan, D. Kumar, Chitosan-based copper nanocomposite accelerates healing in excision wound model in rats, Eur. J. Pharmacol. 731 (2014) 8–19. https://doi.org/10.1016/j.ejphar.2014.02.033

[49]   K.S. Siddiqi, A. ur Rahman, Tajuddin, A. Husen, Properties of Zinc Oxide Nanoparticles and Their Activity Against Microbes, Nanoscale Res. Lett. 13 (2018). https://doi.org/10.1186/s11671-018-2532-3

[50]   I. Kalashnikova, S. Das, S. Seal, Nanomaterials for wound healing: Scope and advancement, Nanomedicine. 10 (2015) 2593–2612. https://doi.org/10.2217/nnm.15.82

[51]   J.C. Chen, B.B. Lin, H.W. Hu, C. Lin, W.Y. Jin, F.B. Zhang, Y.A. Zhu, C.J. Lu, X.J. Wei, R.J. Chen, NGF accelerates cutaneous wound healing by promoting the migration of dermal fibroblasts via the PI3K/Akt-Rac1-JNK and ERK pathways, Biomed Res. Int. 2014 (2014). https://doi.org/10.1155/2014/547187

[52]   I.N. de F. Ramos, M.F. da Silva, J.M.S. Lopes, J.N. Cruz, F.S. Alves, J. de A.R. do Rego, M.L. da Costa, P.P. de Assumpção, D. do S. Barros Brasil, A.S. Khayat, Extraction, Characterization, and Evaluation of the Cytotoxic Activity of Piperine in Its Isolated form and

in Combination with Chemotherapeutics against Gastric Cancer, Molecules. 28 (2023). https://doi.org/10.3390/molecules28145587

[53]   S. BarathManiKanth, K. Kalishwaralal, M. Sriram, S.B.R.K. Pandian, H. seop Youn, S.H. Eom, S. Gurunathan, Anti-oxidant effect of gold nanoparticles restrains hyperglycemic conditions in diabetic mice, J. Nanobiotechnology. 8 (2010) 1–15. https://doi.org/10.1186/1477-3155-8-16

[54]   A. Choudhary, V. Kant, B.L. Jangir, V.G. Joshi, Quercetin loaded chitosan tripolyphosphate nanoparticles accelerated cutaneous wound healing in Wistar rats, Eur. J. Pharmacol. 880 (2020) 173172. https://doi.org/10.1016/j.ejphar.2020.173172

[55]   D.J. Kamothi, V. Kant, B.L. Jangir, V.G. Joshi, M. Ahuja, V. Kumar, Novel preparation of bilirubin-encapsulated pluronic F-127 nanoparticles as a potential biomaterial for wound healing, Eur. J. Pharmacol. 919 (2022) 174809. https://doi.org/10.1016/j.ejphar.2022.174809

[56]   A. Barroso, H. Mestre, A. Ascenso, S. Simões, C. Reis, Nanomaterials in wound healing: From material sciences to wound healing applications, Nano Sel. 1 (2020) 443–460. https://doi.org/10.1002/nano.202000055

[57]   S. Küchler, N.B. Wolf, S. Heilmann, G. Weindl, J. Helfmann, M.M. Yahya, C. Stein, M. Schäfer-Korting, 3D-Wound healing model: Influence of morphine and solid lipid nanoparticles, J. Biotechnol. 148 (2010) 24–30. https://doi.org/10.1016/j.jbiotec.2010.01.001

[58]   Y.H. Lin, J.H. Lin, T.S. Li, S.H. Wang, C.H. Yao, W.Y. Chung, T.H. Ko, Dressing with epigallocatechin gallate nanoparticles for wound regeneration, Wound Repair Regen. 24 (2016) 287–301. https://doi.org/10.1111/wrr.12372

[59]   W. Wang, K.J. Lu, C.H. Yu, Q.L. Huang, Y.Z. Du, Nano-drug delivery systems in wound treatment and skin regeneration, J. Nanobiotechnology. 17 (2019). https://doi.org/10.1186/s12951-019-0514-y

[60]   M. Hajimiri, S. Shahverdi, M.A. Esfandiari, B. Larijani, F. Atyabi, A. Rajabiani, A.R. Dehpour, M. Amini, R. Dinarvand, Preparation of hydrogel embedded polymer-growth factor conjugated nanoparticles as a diabetic wound dressing, Drug Dev. Ind. Pharm. 42 (2016) 707–719. https://doi.org/10.3109/03639045.2015.1075030

[61]   Z. Xie, C.B. Paras, H. Weng, P. Punnakitikashem, L.C. Su, K. Vu, L. Tang, J. Yang, K.T. Nguyen, Dual growth factor releasing multi-functional nanofibers for wound healing, Acta Materialia Inc., 2013. https://doi.org/10.1016/j.actbio.2013.07.030

Nanomaterials in Biological Systems
Materials Research Foundations 185 (2026) 92-111

Materials Research Forum LLC
https://doi.org/10.21741/9781644903858-5

Chapter 5

Nanoparticles Interaction with Cancer Cells

Muhammad Hussnain Siddique[1], Eisha Javed[1], Farrukh Azeem[1], Ijaz Rasul[1], Saima Muzammil[2],
Sumreen Hayat[2], Muhammad Afzal[1], Hamna Maqsood[1], Muhammad Zubair[1]*

[1]Department of Bioinformatics and Biotechnology, Government College University Faisalabad,
Pakistan

[2]Institute of Microbiology, Government College University Faisalabad, Pakistan

muhammad.zubair1751@gmail.com

Abstract

Cancer, a formidable global health concern, is characterized by uncontrolled cell proliferation,
infiltrative potential, and the ominous threat of metastasis. Diverse treatment approaches, from
surgery to immunotherapy, aim to combat this relentless disease. Despite progress, the growing
incidence of cancer demands innovative approaches. Nanotechnology has revolutionized cancer
therapy, enhancing chemotherapy, radiation therapy, and targeted treatments. Nanomaterials,
defined by key attributes, offer early detection and advanced treatment methods. Engineered
nanoparticles deliver precision therapy, triggering cell death or transporting therapeutic agents.
Magnetic nanoparticles, carbon nanotubes, polymeric micelles, and liposomes have found unique
biological applications. In this chapter the interaction between nanoparticles and cancer cells, their
distribution within the body, and impact on the tumor microenvironment are explored.
Furthermore, the chapter addresses the challenges and prospects of this promising approach,
underscoring its potential to reshape the future of cancer therapy.

Keywords

Cancer, Nanotechnology, Nanomaterials, Targeted Therapy, Tumor Microenvironment,
Nanoparticles Uptake, Intracellular Fate of Nanoparticles

Contents

Materials Research Forum LLC
https://doi.org/10.21741/9781644903858-5

1.   Introduction

Cancer is a condition that leads to the uncontrolled proliferation of cells within the body. It's a serious health concern, and it's important to understand its key characteristics, as they play a major role in how it's diagnosed and treated. Cancer is characterized by several key features, including abnormal cell differentiation, rapid cell proliferation, loss of control over cell growth, the capacity of cancer cells to infiltrate neighboring tissues, and the potential to disseminate to distant parts of the body, a process known as metastasis. Each year, the number of new cancer cases continues to rise, this makes it one of the most lethal illnesses worldwide. Cancers can be broadly categorized into several types: leukemia, lymphoma, sarcoma, melanoma, and carcinoma [1]. The treatment of cancer involves various approaches, based on the specific disease type and its current stage of progression. These treatment options encompass surgical interventions, chemotherapy, radiation therapy, targeted therapies, immunotherapy, stem cell or bone marrow transplantation, as well as hormone-based treatments. Surgery is often the first line of treatment and involves the removal of cancerous lesions. In some cases, lymphadenectomy, the removal of lymph nodes, may be performed to enhance the effectiveness of surgery. However, it's important to note that incomplete resection can increase the risk of cancer spreading to other parts of the body. Chemotherapy and radiation therapy are essential treatments that aim to specifically locate and eliminate cancerous cells throughout the body. Targeted therapy, on the other hand, provides a more precise way to transport medications specifically to cancer cells, increasing treatment effectiveness [2]. While there are ongoing research efforts exploring new and emerging therapies, these methods are not yet fully developed for accurate and effective cancer treatment. Nonetheless, significant progress is being made in the fight against cancer, offering hope for improved treatments and outcomes in the future.

Nanotechnology has significantly contributed to the transformation of cancer treatment in recent years. It has had a profound impact on various therapeutic approaches, including chemotherapy, radiation therapy, and targeted therapy. Nanomaterials have become instrumental in the field of nanomedicine, contributing to the development of advanced methods for early cancer diagnosis, detection, and treatment [3]. A lot of factors come into play when considering the potential of these materials in biomedical applications. Numerous elements play a pivotal role in shaping the efficacy and safety of cancer treatments utilizing nanotechnology. These factors encompass characteristics such as porosity, dimension, surface functional groups, electronic features, zeta potential, as well as potential interactions. Collectively, these attributes significantly impact the performance and safety of nanotechnology-driven cancer therapies. Nanoparticles can be specifically engineered to engage with and target cancer cells in various ways, delivering destructive substances or triggering programmed cell death, known as apoptosis. Within this realm, various types of nanoparticles, including magnetic, carbon nanotubes, polymeric micelles, and liposomes, have taken center stage due to their unique biological applications [4]. Their extraordinary chemical, optical, magnetic, and biological properties pave the way for the creation of imaging probes with remarkable qualities.

In this chapter, we will embark on a journey through the scientific landscape of nanoparticles' interactions with cancer cells. We will explore the mechanisms through which nanoparticles enter, target, and affect cancer cells. We will investigate the factors that govern their distribution within the body, their ability to evade the immune system, and their impact on the tumor microenvironment. Additionally, we will discuss the challenges and prospects of this promising approach and the implications for the future of cancer therapy.

2. Nanoparticle Characteristics

A "nanoparticle" is defined as a minuscule particle, with at least one dimension falling within the *size* ranging from 1-100 nanometers. Within this size range, nanoparticles exhibit distinctive physical, chemical, and biological properties that vary significantly from both individual atoms or molecules and bulk materials of the same composition. These nanoparticles can be constructed from a wide array of materials, includes biomolecules, polymers, organics, carbon-based compounds, silicates, non-oxide ceramics, metals, metal oxides, and metals [5].

Moreover, nanoparticles come in various *shapes and forms*, such as spheres, cylinders, platelets, and tubes. Their adaptability is further enhanced through purposeful surface alterations made to satisfy the requirements of the purpose they are intended for. The remarkable variety of NPs, result

Nanomaterials in Biological Systems   Materials Research Forum LLC
Materials Research Foundations 185 (2026) 92-111   https://doi.org/10.21741/9781644903858-5

from factors like their chemical composition, shapes, and surface alterations, combined with the medium they are dispersed in, the state of their distribution, and the wide array of potential surface adjustments, has propelled this field into a dynamic and crucial domain of scientific exploration [6].

One of the most remarkable aspects of the nanoscale is the exceptionally high concentration of atoms or molecules found on the surface of nanoparticles, rather than within the particle itself. This effect is magnified as the particle size decreases. Additionally, the *surface area* available per unit volume of the material becomes exceedingly large in the nanoscale. It is these two characteristics that give rise to the extraordinary physical, chemical, and biological properties of nanomaterials [7]. Some nanoscale materials even exhibit quantum effects, which open the door to a range of intriguing applications [8]. Furthermore, the unique shapes and structures found in nanomaterials, such as carbon nanotubes (CNT) and dendrimers, add an extra dimension to the diversity of applications made possible by these novel materials.

Nanoparticles possess intriguing qualities owing to their small size. These tiny wonders offer fascinating *optical characteristics* that can be finely tuned by adjusting their size and surface properties, impacting the wavelengths of light they absorb or emit. When nanoparticles are smaller than a critical light wavelength, they can even achieve transparency. Moreover, the chemical composition and size of nanoparticles influence their electrical properties, particularly their ionic potential, electron affinity, and electron transport behavior [9].

In the realm of metals, as nanoparticles shrink in size, their sintering and melting temperatures decrease. This makes them excellent candidates for improving *thermal conduction* when incorporated into solid materials. Furthermore, some metals and metal oxides exhibit enhanced magnetic properties as their particle size decreases. In certain cases, individual metallic magnetic nanoparticles can even display a phenomenon known as *superparamagnetic behavior*. These unique properties of nanoparticles hold huge potential for a variety of applications [10].

The significant specific surface area (SA) of nanoparticles underpins a range of their exceptional applications. It boosts *catalytic activity* due to the abundant SA per unit volume and the uniform dispersion of NPs. Their large SA foster robust connections between the NPs and the solid matrix in which they are embedded. In composite materials, various factors like the chemical composition of the nanoparticles, their aspect ratio, the degree of dispersion, and how they interact with the polymer matrix can lead to diverse mechanical properties in the final composites. Notably, this can result in a high elastic modulus without a proportionate reduction in impact strength, a typical issue when using larger particles [11].

Nanoparticles, or NPs, offer exciting potential as versatile tools for both tracking and treating tumors. Their small size and significant surface-to-volume ratio make them well-suited for these purposes. They can be effectively linked to various biomolecules, allowing for *targeted drug delivery*, and enabling anatomical and functional imaging. The process of delivering NPs involves three crucial steps: firstly, ensuring that they are distributed systemically without being sequestered by the reticuloendothelial system (RES); secondly, facilitating their passage through the blood vessels within the tumor; and finally, allowing them to diffuse and enter the cancerous cells [12].

Nanoparticles are especially effective in *fighting cancer* because they tend to accumulate in high concentrations within tumor cells [13]. It's worth noting that designing NPs larger than 50nm can help prevent RES sequestration. This is particularly valuable because the microenvironment of solid tumors is distinct, which includes elements like lymphocytes, extracellular matrix, inflammatory cells, signaling chemicals, and blood vessels. These components create a distinctive landscape within the tumor. In solid tumors, unlike healthy tissues, the microvascular environment is more permeable, featuring capillary pores ranging from 120 to 1200 nanometers in size. This increased permeability allows NPs to penetrate tumors more effectively, making them promising tools for both cancer imaging and drug delivery [14].

3. Types of Nanoparticles Used in Cancer Treatment

Nanomaterials hold great promise for a range of biological applications due to their significant specific surface area, strong surface reactivity, impressive antioxidant capabilities, excellent biocompatibility, and molecular adjustability. In the realm of biomedicine, materials such as liposomes, polymeric micelles, quantum dots, graphene, carbon nanotubes, metallic nanoparticles, and magnetic nanoparticles find common usage. Following are some types of Nanoparticles used in the Cancer Treatment:

3.1 Organic Nanoparticles

Liposomes are double-layered vesicles composed of an enclosed liquid core surrounded by a lipid membrane. This membrane can consist of either naturally occurring or synthetic phospholipids. Liposomes are well-suited for drug delivery due to their amphiphilic properties and their adaptability for surface modifications [15].

Solid lipid nanoparticles (SLNs) represent another type of lipid-based nanoparticles. They are essentially a solid lipid matrix incorporating triglycerides, lipids, fatty acids, steroids, and waxes, typically with a size of less than 1μm. To enhance their stability, surfactants are commonly integrated into their formulation. These nanoparticles can effectively serve as carriers for drugs with poor solubility in aqueous environments. They enable controlled release at specified times and can be administered through various means, such as ingestion and injections, to deliver therapeutic agents to specific target locations [16].

Polymeric nanoparticles are the most used carriers for drug delivery. The polymers used in these nanoparticles can be either natural or synthesized, with a preference for biodegradable options. Polymeric nanoparticles offer significant advantages, including high stability and the potential for mass production. These nanoparticles can be classified into two primary systems: nanocapsules and nanospheres. Nanocapsules store the drug inside a polymer reservoir, while in the case of nanospheres, the drug is evenly distributed within a polymer matrix [17].

3.2 Inorganic Nanoparticles

Gold Nanoparticles

Gold nanoparticles (AuNPs) have garnered significant concentration in the realm of biological applications, owing to their numerous advantages. These versatile Au NPs come in various forms, ranging from spheres to rods, cages, and more, with diameters ranging from 1nm to 100nm. Their shape and size play a crucial role in determining their optical and electrical properties. One key attribute of AuNPs is their negative charge, making them readily responsive to functionalization with a wide array of biomolecules, including genes and targeting ligands. Notably, AuNPs are both safe and biocompatible, making them well-suited for various biological applications. These nanoparticles exhibit surface plasmon resonance (SPR) bands, highlighting their unique qualities, including a distinct surface effect, a macroscopic quantum tunnelling effect, and ultra-small size. Thanks to these exceptional properties, AuNPs have appeared as a highly promising material for a range of biological endeavors, for instance facilitating drug delivery, enabling molecular imaging, and enhancing biosensing techniques.

Properties such as absorption and scattering of radiation make gold NPs interesting for assistance in photothermal therapy (PTT). In PTT, electromagnetic radiation is utilized to produce heat, which, in turn, is employed to selectively destroy cancer cells [18].

Silver Nanoparticles

Utilizing silver nanoparticles (AgNPs) as a harmless and effective method for cancer treatment presents unique challenges. Silver nanoparticles (AgNPs) provide a diverse set of benefits owing to their properties that are influenced by their size and shape. These advantages encompass a wide spectrum of features, including magnetic, optical, chemical, and physical characteristics. These versatile AgNPs find applications in various products, including biosensors, antimicrobial agents, cosmetics, and electrical components. They are also employed in fields like medical imaging, drug delivery, and cell electrodes. What sets silver NPs apart is their exceptional light absorption, high resolution, and strong affinity for functionalization. In addition to their ability to hinder the growth of cancer cells, these nanoparticles have the potential to trigger pathways that actively suppress the process of cell division. This makes them valuable for detecting various cancer, such as lung, prostate, hepatic, and cervical cancer [19].

Palladium Nanoparticles

The emergence of palladium nanoparticles (PdNPs) represents a noteworthy leap in the application of noble metal NPs. These PdNPs offer significant benefits when compared to their noble metal counterparts, boasting biocompatibility and excellent photothermal stability. These qualities position PdNPs as promising candidates in the field of biomedicine, particularly in the context of multimodal imaging-guided cancer treatment. PdNPs exhibit impressive attributes, including exceptional near-infrared (NIR) absorption, rapid photothermal conversion rates, and remarkable biocompatibility. Variants such as Pd nanosheets (PdNSs), porous/hollow PdNPs, and Pd-based hybrids like Pd@M (where M can stand for Ag, Pt, ZIF-8, SiO_2) exemplify this family of nanomaterials. These Pd-based nanoparticles have garnered attention for their potential use as both therapeutic agents and contrast agents in the domain of cancer imaging. The optical absorption peaks of Pd nanosheets differ with their size but consistently fall within the NIR range. This quality holds promise for their application in photothermal therapy (PTT), a significant development in the cancer therapy [20].

Iron Oxide Nanoparticles

Iron oxide nanoparticles are highly valued in biological applications thanks to their remarkable magnetic properties, well-suited ratio of surface to volume for functionalization, and compatibility with living systems. These nanoparticles have gained a significant grip in the medical field, finding use as contrast agents and mediators for heating in techniques like magnetic resonance imaging (MRI) and magnetic particle hyperthermia. Superparamagnetic iron oxide nanoparticles (SPIONs) have emerged as a focal point of interest due to their exceptional superparamagnetic behavior, ability to generate heat through magnetic fields, and their capacity to enhance magnetic resonance imaging. When combined with medications, SPIONs bring about advantages like *in vivo* imaging, magnetic thermotherapy, and the concurrent administration of anti-cancer medications, making them valuable components of advanced drug delivery nano systems [18, 21].

Copper Nanoparticles

Copper-based nanoparticles have garnered significant attention in biomedical applications. These nanoparticles exhibit outstanding near-infrared light absorption when exposed to a near-infrared laser. They excel at converting this absorbed light into heat, which can be used to selectively destroy tumors. Smaller copper NPs also offer the advantage of emitting fluorescence signals and serving as effective tools for optical imaging. Furthermore, CuNPs present a useful platform for drug delivery and image-guided treatments. Their advantages over gold (Au) and silver (Ag) NPs include cost-effectiveness, enhanced cytotoxicity towards cancerous cells at lower doses, and extended stability [22].

Notably, innovative copper-containing nanoparticles, like copper selenide nanocrystals, nanocubes, monodispersed Cu-Te nanorods, nanoplates, and copper bismuth sulfide nanostructures, have been engineered and verified for their potential in photothermal therapy (PTT). In summary, these copper-based nanomaterials show strong promise in the fight against cancer, with improved efficiency in generating heat for therapeutic purposes [23].

Table 1: List of different types of NPs for Cancer Therapy

Nanoparticle Type	Composition	Targeted Therapy	Advantages
Liposomes	Lipid bilayers	Drug delivery, gene therapy	Enhanced drug stability, controlled release
Polymeric Nanoparticles	Biodegradable polymers (e.g., PLGA, PEG)	Chemotherapy, immunotherapy	Customizable drug release profiles, high drug loading capacity
Metal Nanoparticles	Gold, silver, iron oxide	Photothermal therapy, targeted drug delivery	Strong light absorption, potential for imaging and therapy
Dendrimers	Branched macromolecules	Drug delivery, gene therapy	Multivalency for targeted delivery, precise size control
Silica Nanoparticles	Silicon dioxide	Drug delivery, imaging	High surface area for drug loading, easy surface modification
Carbon Nanotubes	Cylindrical carbon structures	Drug delivery, photothermal therapy	High surface area, excellent thermal conductivity
Polymeric Micelles	Amphiphilic block copolymers	Drug delivery, targeted therapy	Enhanced drug solubility, active targeting capabilities
Lipid Nanoparticles	Lipids, phospholipids	mRNA delivery, gene therapy	Efficient nucleic acid encapsulation, low immunogenicity
Nanogels	Crosslinked polymer networks	Drug delivery, radiotherapy	High water content, responsiveness to stimuli

Quantum Dots Nanoparticles

Quantum dots (QDs) are semiconductor nanoparticles known for their exceptional optical characteristics resulting from quantum and size effects. These unique properties are primarily attributed to inorganic materials like CdS, CdTe, ZnS, and PbS. Among these, the most widely employed QD configuration consists of a semiconductor core made from CdSe, enclosed within an outer shell of ZnS [16, 24].

Nanomaterials in Biological Systems   Materials Research Forum LLC
Materials Research Foundations 185 (2026) 92-111   https://doi.org/10.21741/9781644903858-5

Because of their distinctive attributes, such as resistance to photobleaching, sustained and intense fluorescence over extended periods, and heightened sensitivity in detection, quantum dots have emerged as a groundbreaking class of biosensors. They find significant utility in the realm of cancer diagnosis, offering advanced capabilities for precise and sensitive detection in this critical area of healthcare. *Table 1* presents different types of NPs that are used for cancer therapy.

4.   Cancer Cell Characteristics

The concept of the Hallmarks of Cancer represents a framework that describes the functional attributes acquired by human cells during their transition from a state of normalcy to the development of malignant tumors. In essence, these hallmarks encapsulate the critical capabilities enabling the formation of cancerous growths. At present, there are eight identified hallmarks, each representing a unique trait crucial for tumor malignancy. These abilities encompass the potential to maintain continuous signaling for growth, elude mechanisms that inhibit proliferation, withstand cellular apoptosis (programmed cell death), attain the capability for unlimited replication, trigger the process of angiogenesis (the formation of blood vessels), activate pathways for invasive behavior and metastatic spread, reconfigure cellular metabolism, and avoid immune system-mediated destruction [25].

These hallmarks collectively constitute a comprehensive framework for understanding the complex and multifaceted nature of cancer, shedding light on the multitude of factors and processes that contribute to the emergence and progression of this disease. All these factors are collectively called the characteristics of cancer cells that help them grow under the complex immune system.

5.   Challenges in Targeting Cancer Cells

Despite the numerous therapeutic options available to cancer patients, the formidable obstacle of drug resistance persists as a significant challenge. This challenge is especially pronounced in the context of metastatic disease, which often remains incurable due to the remarkable resilience of metastatic cells to current treatments. A growing body of evidence underscores the presence of a small subgroup of cells within tumors, known as cancer stem cells (CSCs), that possess the unique ability to initiate the formation of primary tumors and facilitate the spread of cancer to distant sites, a process known as metastasis. CSCs possess a repertoire of mechanisms that confer resistance to various treatments. These include robust DNA damage repair mechanisms, activation of survival pathways, heightened cellular adaptability, immune evasion strategies, and the capacity to thrive in hostile microenvironments [26].

Cancerous tumors are characterized by genetic heterogeneity, meaning that cancer cells in a tumor often exhibit varying genetic mutations. This diversity presents a significant challenge in the quest for the next generation of targeted anticancer drugs. Furthermore, current research has underscored the critical role of the tumor microenvironment (TME) in shaping the development and testing of novel cancer treatments [27]. TME encompasses all the elements within a tumor, including a diverse array of genetically distinct red blood cells, white blood cells, platelets, healthy normal cells, cancerous cells, endothelial cells. Importantly, various cell types within the TME play a pivotal role in promoting cancer progression. Notably, cancer-associated fibroblasts (CAFs), tumor-associated macrophages, and endothelial cells have been identified as key contributors to tumor development and aggressivity across various cancer types [28].

Nanomaterials in Biological Systems   Materials Research Forum LLC
Materials Research Foundations 185 (2026) 92-111   https://doi.org/10.21741/9781644903858-5

The presence of distinct cell populations within a tumor presents a major challenge in the field of cancer research, demanding an approach that addresses both intra-tumoral and inter-tumoral heterogeneity [29]. Consequently, the pursuit of targeting the molecular drivers active within CSCs, in conjunction with standard treatments, holds the potential to enhance treatment outcomes for cancer patients. This integrated strategy aims to achieve enduring responses and a more favorable prognosis for individuals battling this complex and relentless disease. Understanding and effectively targeting both the genetic heterogeneity of cancer cells and the complex interactions within the TME are critical avenues in the ongoing pursuit of innovative cancer therapies [30].

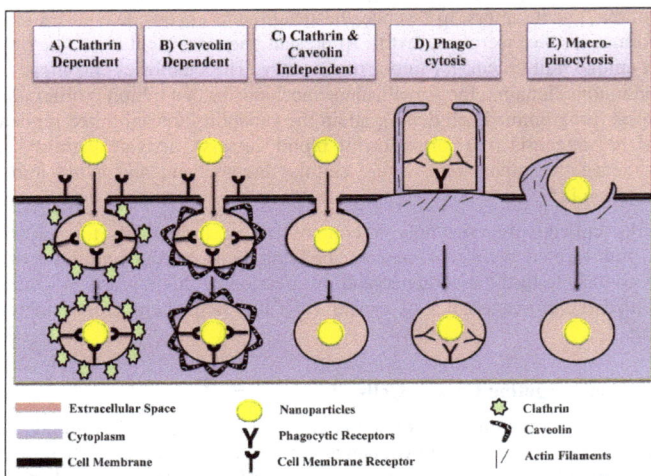

Figure 1. Schematic representation of nanoparticle uptake pathways involving endocytosis. Diverse mechanisms are present for the cellular entry of NPs through endocytosis, including (A) Clathrin-dependent; (B) Caveolin-dependent; (C) Clathrin- and caveolin-independent; (D) Phagocytosis; and (E) Macropinocytosis pathways. These pathways for NPs cellular uptake exhibit distinct mechanistic processes, tightly controlled at the biomolecular level. The chosen pathway for nanoparticle entry plays a pivotal role, influencing intracellular NPs transport and subsequently affecting biological responses and therapeutic outcomes.

6.   Mechanisms of Nanoparticle Uptake

In the context of cellular metabolism, macromolecules and nanomaterials typically cannot directly pass through the cellular lipid bilayer. Nanoparticles have various pathways to traverse a cell's plasma membrane, whether during *in vivo* or *in vitro* exposure. These pathways can be broadly classified into two main groups: *(i)* Uptake pathways based on endocytosis (*Fig. 1*); and *(ii)* Direct entry of nanoparticles into the cell (*Fig. 2*) [26]. These nanomaterials, once endocytosed, become confined within membrane-bound vesicles known as endolysosomes and are unable to access the cell's cytosol. This energy-dependent endocytosis is often referred to as vesicle transport. In the study of intracellular trafficking of nanomaterials, endocytosis can be categorized into two primary types: phagocytosis and pinocytosis. Phagocytosis refers to the solid-uptake pathways observed in processes like those in bacteria and yeast, while pinocytosis describes the fluid-phase uptake pathway. These distinctions help us understand how different nanomaterials are taken up by cells [31].

When nanoparticles are intended for intravenous administration with the goal of accumulating in tumors, they are often coated with PEG layers. Whether it's a liposome, dendrimer, or inorganic nanoparticle, this layer, known as a "corona" or "halo," serves to extend their circulation time in the bloodstream [32]. The PEG coating effectively hinders the binding of opsonins and their uptake by macrophages, enhancing the nanoparticles' ability to reach and accumulate within tumor sites.

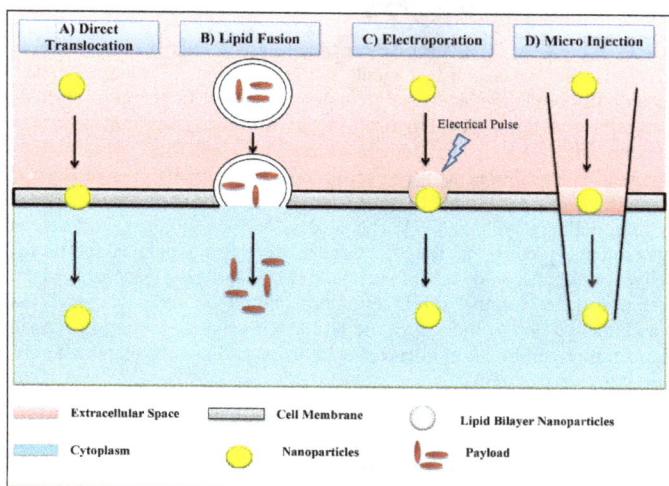

Figure 2. Illustrative overview of methods and approaches for NPs to achieve direct cytoplasmic delivery by traversing the cell plasma membrane. These include: (A) Direct translocation; (B) Lipid fusion; (C) Electroporation; and (D) Microinjection. Each of these paths facilitates the direct entry of nanoparticles into the cell's cytoplasm. The physicochemical characteristics of the nanoparticles affect direct translocation and lipid fusion; electroporation uses electrical pulses to damage the cell plasma membrane; and microinjection employs a microscopic needle to puncture the plasma membrane, injecting nanoparticles directly into the cytoplasm.

7. Targeting of Cancer Cells by Nanoparticles

Copper nanoparticles have emerged as a promising avenue in the pursuit of potential anticancer therapies. Their remarkable capacity to trigger oxidative stress in cancer cells through the production of reactive oxygen species (ROS), ultimately causing cell death, is particularly noteworthy. These nanoparticles have demonstrated a capacity to particularly focus on cancer cells while preserving healthy cells. Moreover, they have been found to enhance the sensitivity of tumor cells to radiation therapy, ultimately contributing to tumor reduction [33].

Silver nanoparticles have also garnered attention for their potential in cancer treatment. They exert their effects via interacting with cell surface proteins, inducing cell death through ROS production, and exhibiting a remarkable ability to attack just cancerous cells. Furthermore, AgNPs have shown promise as imaging agents in cancer identification, owing to their potent optical properties [34].

Zinc oxide nanoparticles are another intriguing option under investigation for their anticancer potential. These nanoparticles can trigger apoptosis in cancer cells, targeting them precisely and activating caspases and suppressing anti-apoptotic proteins to cause cell death. Additionally, they have demonstrated the capacity to reduce cancer cell drug resistance, thus enhancing the effectiveness of chemotherapy [35].

Ongoing research in the field of NPs for cancer targeting is primarily centered on enhancing the precision and effectiveness of established agents, while exploring novel nanoparticle varieties with distinct characteristics for the diagnosis and treatment of cancer. In parallel, substantial efforts are underway in research and development to discover and formulate drugs that interfere with or block signal transduction pathways and are unique to cancer cells [36]. These drugs hold the promise of personalized cancer care, tailored to the specific molecular targets presented by each patient's tumor. They can be administered directly to the tumor's location or the cancer cell surface, often facilitated by drug delivery systems [37]. The success of such targeted drug delivery hinges on three key components: identifying specific targets, developing ligands for these targets, and employing delivery systems conjugated to the ligands [38]. Tailoring these strategies to the diverse nature of different cancers, whether solid tumors or hematologic malignancies, is essential. Solid tumors pose a challenge due to their dynamic and heterogeneous biology, necessitating a deep understanding of tumor cell biology, microenvironments, and growth patterns to create effective and targeted drug delivery systems.

Targeted drug delivery can be achieved through two principal approaches: actively engaging in the search for drug carriers featuring ligands designed for specific targeting or capitalizing on the unique pathophysiological characteristics of tumor tissue [39]. Cryosurgery, a procedure involving the freezing of tumor tissue, offers several advantages, including cost-effectiveness, reduced invasiveness, minimal intraoperative bleeding, and fewer postoperative complications. However, it also has limitations, such as inefficient freezing and potential damage to adjacent tissues [40]. To address these challenges, researchers have explored the use of antifreeze proteins like AFP-1 to improve cryosurgery outcomes, though results have been somewhat mixed. With the advancement of nanotechnology, a novel approach called nano-cryosurgery has emerged. *Nano-cryosurgery* hinges on the precise delivery of NPs with specific physiochemical properties into tumor tissues. Leveraging the unique characteristics of nanomaterials, this technique allows for better control of the freezing process, including the alteration of the range, creation of ice balls, and enhancement of freezing effectiveness [41]. Consequently, nano-cryosurgery can effectively eliminate cancerous tissue while safeguarding neighboring normal tissue from freezing damage, representing a promising development In the realm of cancer therapy.

Nanomaterials in Biological Systems   Materials Research Forum LLC
Materials Research Foundations 185 (2026) 92-111   https://doi.org/10.21741/9781644903858-5

8.   Intracellular Fate of Nanoparticles

Nanoparticles that rely on *phagocytosis* to enter cells undergo a complex process. Initially, they must be recognized by opsonins like Antibodies such as IgG and IgM, alongside elements of the complement system including C3, C4, and C5, as well as various proteins found in blood serum [42]. Once opsonized, these NPs adhere to the cellular membrane, triggering a cup-shaped membrane extension. This extension eventually surrounds and internalizes the nanoparticles, forming phagosomes with diameters ranging from 0.5 to 10µm. Finally, phagosomes migrate towards the process of merging with lysosomes, where the cargo enclosed within them is broken down through acidification and enzymatic processes [43]. To achieve desired effects, nanomedicines need to circumvent this route to evade degradation.

Clathrin-dependent endocytosis is a prevalent mechanism of cellular entry that involves the interaction of nanomaterials with specific receptors on the cell surface. Clathrin proteins polymerize on the cytosolic side of the plasma membrane, forming clathrin-coated vesicles (CCVs) that encase the nanoparticles [44]. These vesicles are then pinched off with the help of dynamin, driven by energy from actin. CCVs move within cells, guided by the cytoskeleton, until they shed their clathrin coat. The fate of these vesicles depends on the receptors to which the nanoparticles' ligands attach. For example, they may be transported to lysosomes for degradation or returned to the surface of the cell [45]. Inhibitors or elements like chlorpromazine, a hypertonic medium, or decreased potassium levels may have an impact on this pathway.

Caveolae-dependent endocytosis is another frequent route for cellular entrance. It enables nanoparticles to bypass lysosomal degradation, making it a preferred route for enhancing targeting precision and therapeutic efficacy [46]. In this process, caveolin, a protein found in maximum cells, serves a crucial part. Nanoparticles or pathogens interact with plasma membrane receptors, inducing the development of flask-shaped vesicles that dynamin pinches away from the membrane. These caveolar vesicles travel to merge with caveosomes or multivesicular bodies (MTV) with a neutral pH [47]. Subsequently caveosomes carrying nanomedicine travel along microtubules to the endoplasmic reticulum [48]. Here, it is believed that NPs pass through the cytosol before reaching the nucleus using the nuclear pore complex. Although it takes longer, this method produces smaller vesicles than clathrin-dependent endocytosis. Choosing this pathway can help nanoparticles avoid degradation and improve delivery to a specific organelle, such as the ER or nucleus, crucial for improved therapeutic delivery [49].

Macropinocytosis is a widespread cellular mechanism, initiated by external stimuli activating receptor tyrosine kinases. This process results in the internalization of the surrounding fluid into large vacuoles [50]. The nature of the ruffles formed on the cell membrane during this process determines the mechanism by which macropinosomes are pinched off. Dynamin trims circular ruffles, while lamellipodial macropinosomes are split without dynamin involvement. Macropinosomes, typically ranging from 0.5 to 10µm in diameter, distinguish themselves from other vesicles formed in different pinocytosis processes. These vesicles may either fuse with lysosomes in macrophages or return to the cell surface in certain cell types like human A431 cells, releasing their contents into the extracellular space. The outcome of macropinosomes is determined by the specific type of cell in which they are found [51]. *Table 2* represents some approaches of NPs for targeting various cellular organelles.

Table 2: Approaches of NPs for Targeting Various Cellular Organelles

Targeted Organelle	Targeting Strategies	Nanomaterials	Active/Passive Targeting
Nucleus	Developing nuclear pore complex penetrating nanomaterials	Au-tiopronin-triplex-forming oligonucleotides, nucleolin-specific aptamers, gold-silver nanorod, Tat peptide, etc.	Passive
Mitochondria	Utilizing mitochondrial membrane potential and import machinery	Platinum-based therapeutic agents, multi-layered carbon nanotubes, rhodamine-110 dye, CoQ10, and polymeric PEG compounds, etc.	Active
Endoplasmic Reticulum	Designing ER-specific structures	PEG-PE Micelles, Thermoresponsive Micelles, PU-ER Liposomes, ERL Liposome, Antigen Peptide Nanoparticles, PGA-based Nanoparticles, etc.	Passive
Lysosome	Designing pH-responsive nanoparticles	DOX-Polymers, DSPE-MPEG-Se, self-assembled Squalen-Penicil bioconjugates, etc.	Passive

9. Challenges and Limitations

Exploring the interaction of nanoparticles with cancer cells represents a promising frontier in the battle against cancer. However, like any emerging field, it faces various challenges and limitations. Understanding and addressing these obstacles is crucial for the advancement of nanoparticle-based cancer therapies. Here are some of the key challenges and limitations in this area:

9.1 Heterogeneity of Cancer Cells

Cancer is a highly heterogeneous disease, with variations in cell types, genetic mutations, and behaviors. Developing nanoparticles that can effectively target and interact with this diverse array of cancer cells is a complex challenge [52].

9.2 Limited Penetration

Nanoparticles often struggle to penetrate deep into solid tumor tissues. Their ability to navigate through the dense extracellular matrix and reach the core of the tumor is limited. This can reduce their therapeutic effectiveness, especially in larger or more solid tumors [53].

9.3 Immune System Recognition

The immune system may recognize nanoparticles as foreign entities, leading to clearance from the bloodstream. Achieving prolonged circulation is essential for effective therapy, and evading immune responses can be challenging [54].

9.4 Toxicity Concerns

Some nanoparticles can be toxic to healthy tissues. Understanding and mitigating potential toxic effects, especially when using non-biodegradable materials, is a significant concern [55].

Nanomaterials in Biological Systems   Materials Research Forum LLC
Materials Research Foundations 185 (2026) 92-111   https://doi.org/10.21741/9781644903858-5

9.5   Drug Resistance

Cancer cells can develop resistance to therapeutic agents, including those delivered by nanoparticles. Overcoming or circumventing these resistance mechanisms is a constant challenge [56].

9.6   Off-Target Effects

Even when designed for targeted delivery, nanoparticles may exhibit off-target effects, causing unintended harm to healthy cells and tissues. Achieving high specificity in targeting cancer cells while sparing normal ones remains a challenge [53].

9.7   Regulatory and Clinical Hurdles

Transitioning nanoparticle-based therapies from the lab to the clinic involves navigating complex regulatory and clinical trial processes. Demonstrating safety and efficacy through rigorous testing and approval procedures is resource-intensive and time-consuming [57].

9.8   Patient Variability

Patients exhibit variations in their response to nanoparticle-based therapies due to differences in genetics, metabolism, and tumor characteristics. Personalized treatment approaches may be required to account for this variability [58].

9.9   Ethical and Societal Concerns

The use of nanotechnology in medicine raises ethical and societal questions about safety, privacy, and equitable access to advanced treatments. Addressing these concerns is crucial for responsible development [59].

9.10   Long-Term Effects

The long-term consequences of NPs in the body are not completely understood. Concerns about possible accumulative toxicity and delayed side effects require ongoing investigation [60].

9.11   Interdisciplinary Collaboration

Progress in this field often depends on interdisciplinary collaboration between researchers, clinicians, engineers, and regulatory bodies. Bridging the gap between different areas of expertise can be a challenge [61].

While the interaction between nanoparticles and cancer cells holds great potential, these challenges and limitations emphasize the need for ongoing research and innovation. Researchers are continually working to address these issues and develop safe and effective nanoparticle-based cancer therapies that can make a meaningful impact on cancer treatment and patient outcomes.

10.   Future Perspectives

The future perspectives for the interaction of nanoparticles with cancer cells is exceptionally promising, with continued research and development opening up exciting possibilities and avenues for exploration:

10.1 Personalized Medicine

The future of cancer therapy is likely to be increasingly personalized. Nanoparticle-based approaches can be tailored to individual patients, considering their genetic makeup, tumor characteristics, and treatment responses. This could lead to more effective and less toxic treatments.

10.2 Immunotherapy Integration

Combining nanoparticle-based drug delivery with immunotherapies holds great potential. Nanoparticles can help modulate the tumor microenvironment, making it more receptive to immune interventions and enhancing the overall effectiveness of cancer immunotherapies.

10.3 Multimodal Imaging and Therapy

Nanoparticles with integrated imaging capabilities will enable real-time monitoring of treatment responses. This allows for quick adjustments to therapy and the potential for on-the-fly personalized treatment regimens.

10.4 Reducing Drug Resistance

Ongoing research will aim to overcome the challenges of drug resistance by developing innovative nanoparticle strategies that can circumvent or counter resistance mechanisms employed by cancer cells.

10.5 Bioinformatics and Data Integration

Advanced data analytics and machine learning techniques will play a pivotal role in understanding patient responses to nanoparticle-based therapies. This will aid in treatment optimization and patient stratification.

10.6 Emerging Nanomaterials

The discovery and development of novel nanomaterials with improved properties, such as increased drug loading capacity or enhanced targeting, will continue to expand the possibilities in cancer therapy.

The future of nanoparticles interacting with cancer cells is both promising and dynamic, holding the capability to revolutionize cancer therapy and enhance patient outcomes. Ongoing research and new discoveries are steering the field towards the development of safe, efficient, and patient-centric nanoparticle-based cancer therapies.

11. Summary

The field of nanoparticles and their interaction with cancer cells holds immense promise for the future of cancer treatment. The burden of cancer is substantial, and innovative approaches are crucial. Nanoparticles, with their unique attributes, offer a potential solution. Size, shape, surface properties, and material composition determine their efficacy in targeting cancer cells. Distinct types of nanoparticles, such as liposomes, solid lipid nanoparticles, silver NPs, quantum dot NPs, and gold nanoparticles, present varied advantages, and limitations. The challenge lies in the heterogeneity and adaptability of cancer cells, complicating precise targeting. The hurdles include the complex tumor environment, drug delivery issues, and resistance mechanisms. Understanding how nanoparticles enter cancer cells sheds light on their potential. Active and passive targeting strategies are essential, as is comprehending the intracellular journey of nanoparticles. However, concerns like toxicity, off-target effects, and aggregation must be addressed with rigorous safety assessments. Looking forward, ongoing research and innovative technologies can overcome existing limitations, potentially revolutionizing cancer therapy. Precision medicine and advanced imaging hold promise in tailoring nanoparticle treatments for individual patients. Collaboration among researchers, clinicians, and industry leaders is key to transforming these findings into effective cancer treatments. The future is bright in the fight against cancer.

References

[1] K.A. Oien, Pathologic Evaluation of Unknown Primary Cancer, Semin. Oncol. 36 (2009) 8–37. https://doi.org/10.1053/j.seminoncol.2008.10.009

[2] P.S. Roy, B.J. Saikia, Cancer and cure: A critical analysis, Indian J. Cancer. 53 (2016) 441–442. https://doi.org/10.4103/0019-509X.200658

[3] O. Adir, M. Poley, G. Chen, S. Froim, N. Krinsky, J. Shklover, J. Shainsky-Roitman, T. Lammers, A. Schroeder, Integrating Artificial Intelligence and Nanotechnology for Precision Cancer Medicine, Adv. Mater. 32 (2020). https://doi.org/10.1002/adma.201901989

[4] M.M. Rahman, M.R. Islam, S. Akash, M. Harun-Or-Rashid, T.K. Ray, M.S. Rahaman, M. Islam, F. Anika, M.K. Hosain, F.I. Aovi, H.A. Hemeg, A. Rauf, P. Wilairatana, Recent advancements of nanoparticles application in cancer and neurodegenerative disorders: At a glance, Biomed. Pharmacother. 153 (2022) 113305. https://doi.org/10.1016/j.biopha.2022.113305

[5] J.N. Cruz, S. Muzammil, A. Ashraf, M.U. Ijaz, M.H. Siddique, R. Abbas, M. Sadia, Saba, S. Hayat, R.R. Lima, A review on mycogenic metallic nanoparticles and their potential role as antioxidant, antibiofilm and quorum quenching agents, Heliyon. 10 (2024). https://doi.org/10.1016/j.heliyon.2024.e29500

[6] A.L.B. de Barros, A. Tsourkas, B. Saboury, V.N. Cardoso, A. Alavi, Emerging role of radiolabeled nanoparticles as an effective diagnostic technique, EJNMMI Res. 2 (2012) 1–15. https://doi.org/10.1186/2191-219X-2-39

[7] C. Negin, S. Ali, Q. Xie, Application of nanotechnology for enhancing oil recovery – A review, Petroleum. 2 (2016) 324–333. https://doi.org/10.1016/j.petlm.2016.10.002

[8] A. Ike Onyia, H. Ifeanyi Ikeri, A. Iheanyichukwu Chima, Surface and Quantum Effects in Nanosized Semiconductor, Am. J. Nano Res. Appl. 8 (2020) 35. https://doi.org/10.11648/j.nano.20200803.11

[9] E.C. Cho, C. Glaus, J. Chen, M.J. Welch, Y. Xia, Inorganic nanoparticle-based contrast agents for molecular imaging, Trends Mol. Med. 16 (2010) 561–573. https://doi.org/10.1016/j.molmed.2010.09.004

[10] Nanoparticles, Nanostructure Sci. Technol. (2004). https://doi.org/10.1007/978-1-4419-9042-6

[11] M.H. Al-Saleh, U. Sundararaj, Review of the mechanical properties of carbon nanofiber/polymer composites, Compos. Part A Appl. Sci. Manuf. 42 (2011) 2126–2142. https://doi.org/10.1016/j.compositesa.2011.08.005

[12] O. Veiseh, J.W. Gunn, M. Zhang, Design and fabrication of magnetic nanoparticles for targeted drug delivery and imaging, Adv. Drug Deliv. Rev. 62 (2010) 284–304. https://doi.org/10.1016/j.addr.2009.11.002

[13] T. Sun, Y.S. Zhang, B. Pang, D.C. Hyun, M. Yang, Y. Xia, Engineered nanoparticles for drug delivery in cancer therapy, Angew. Chemie - Int. Ed. 53 (2014) 12320–12364. https://doi.org/10.1002/anie.201403036

[14] Y. Dai, C. Xu, X. Sun, X. Chen, Nanoparticle design strategies for enhanced anticancer therapy by exploiting the tumour microenvironment, Chem. Soc. Rev. 46 (2017) 3830–3852. https://doi.org/10.1039/c6cs00592f

[15] S. Fathi-karkan, R. Arshad, A. Rahdar, A. Ramezani, R. Behzadmehr, S. Ghotekar, S. Pandey, Recent advancements in the targeted delivery of etoposide nanomedicine for cancer therapy: A comprehensive review, Eur. J. Med. Chem. 259 (2023) 115676. https://doi.org/10.1016/j.ejmech.2023.115676

[16] A. Aghebati-Maleki, S. Dolati, M. Ahmadi, A. Baghbanzhadeh, M. Asadi, A. Fotouhi, M. Yousefi, L. Aghebati-Maleki, Nanoparticles and cancer therapy: Perspectives for application of nanoparticles in the treatment of cancers, J. Cell. Physiol. 235 (2020) 1962–1972. https://doi.org/10.1002/jcp.29126

[17] S. Parveen, S.K. Sahoo, Polymeric nanoparticles for cancer therapy, J. Drug Target. 16 (2008) 108–123. https://doi.org/10.1080/10611860701794353

[18] A. Pugazhendhi, T.N.J.I. Edison, I. Karuppusamy, B. Kathirvel, Inorganic nanoparticles: A potential cancer therapy for human welfare, Int. J. Pharm. 539 (2018) 104–111. https://doi.org/10.1016/j.ijpharm.2018.01.034

[19] F.S. Alves, J.N. Cruz, I.N. de Farias Ramos, D.L. do Nascimento Brandão, R.N. Queiroz, G.V. da Silva, G.V. da Silva, M.F. Dolabela, M.L. da Costa, A.S. Khayat, J. de Arimatéia Rodrigues do Rego, D. do Socorro Barros Brasil, Evaluation of Antimicrobial Activity and Cytotoxicity Effects of Extracts of Piper nigrum L. and Piperine, Separations. 10 (2023). https://doi.org/10.3390/separations10010021

[20] D. MubarakAli, H. Kim, P.S. Venkatesh, J.W. Kim, S.Y. Lee, A Systemic Review on the Synthesis, Characterization, and Applications of Palladium Nanoparticles in Biomedicine, Appl. Biochem. Biotechnol. 195 (2023) 3699–3718. https://doi.org/10.1007/s12010-022-03840-9

[21] M.P. Vinardell, M. Mitjans, Antitumor activities of metal oxide nanoparticles, Nanomaterials. 5 (2015) 1004–1021. https://doi.org/10.3390/nano5021004

[22] I.N. de F. Ramos, M.F. da Silva, J.M.S. Lopes, J.N. Cruz, F.S. Alves, J. de A.R. do Rego, M.L. da Costa, P.P. de Assumpção, D. do S. Barros Brasil, A.S. Khayat, Extraction, Characterization, and Evaluation of the Cytotoxic Activity of Piperine in Its Isolated form and in Combination with Chemotherapeutics against Gastric Cancer, Molecules. 28 (2023). https://doi.org/10.3390/molecules28145587

[23] L. Li, H. Liu, Biodegradable inorganic nanoparticles: An opportunity for improved cancer therapy?, Nanomedicine. 12 (2017) 959–961. https://doi.org/10.2217/nnm-2017-0057

[24] C.E. Probst, P. Zrazhevskiy, V. Bagalkot, X. Gao, Quantum dots as a platform for nanoparticle drug delivery vehicle design, Adv. Drug Deliv. Rev. 65 (2013) 703–718. https://doi.org/10.1016/j.addr.2012.09.036

[25] D. Hanahan, Hallmarks of Cancer: New Dimensions, Cancer Discov. 12 (2022) 31–46. https://doi.org/10.1158/2159-8290.CD-21-1059

[26] A. Turdo, V. Veschi, M. Gaggianesi, A. Chinnici, P. Bianca, M. Todaro, G. Stassi, Meeting the challenge of targeting cancer stem cells, Front. Cell Dev. Biol. 7 (2019). https://doi.org/10.3389/fcell.2019.00016

[27] S.S. Linton, S.G. Sherwood, K.C. Drews, M. Kester, Targeting cancer cells in the tumor microenvironment: Opportunities and challenges in combinatorial nanomedicine, Wiley Interdiscip. Rev. Nanomedicine Nanobiotechnology. 8 (2016) 208–222. https://doi.org/10.1002/wnan.1358

[28] M.H. Sarfraz, M. Zubair, B. Aslam, A. Ashraf, M.H. Siddique, S. Hayat, J.N. Cruz, S. Muzammil, M. Khurshid, M.F. Sarfraz, A. Hashem, T.M. Dawoud, G.D. Avila-Quezada, E.F. Abd_Allah, Comparative analysis of phyto-fabricated chitosan, copper oxide, and chitosan-based CuO nanoparticles: antibacterial potential against Acinetobacter baumannii isolates and anticancer activity against HepG2 cell lines, Front. Microbiol. 14 (2023). https://doi.org/10.3389/fmicb.2023.1188743

[29] J. Zugazagoitia, C. Guedes, S. Ponce, I. Ferrer, S. Molina-Pinelo, L. Paz-Ares, Current Challenges in Cancer Treatment, Clin. Ther. 38 (2016) 1551–1566. https://doi.org/10.1016/j.clinthera.2016.03.026

[30] A. Agliano, A. Calvo, C. Box, The challenge of targeting cancer stem cells to halt metastasis, Semin. Cancer Biol. 44 (2017) 25–42. https://doi.org/10.1016/j.semcancer.2017.03.003

[31] K. Kettler, K. Veltman, D. van de Meent, A. van Wezel, A.J. Hendriks, Cellular uptake of nanoparticles as determined by particle properties, experimental conditions, and cell type, Environ. Toxicol. Chem. 33 (2014) 481–492. https://doi.org/10.1002/etc.2470

[32] M. Wang, M. Thanou, Targeting nanoparticles to cancer, Pharmacol. Res. 62 (2010) 90–99. https://doi.org/10.1016/j.phrs.2010.03.005

[33] J.J. Xu, W.C. Zhang, Y.W. Guo, X.Y. Chen, Y.N. Zhang, Metal nanoparticles as a promising technology in targeted cancer treatment, Drug Deliv. 29 (2022) 664–678. https://doi.org/10.1080/10717544.2022.2039804

[34] Neha Desai, M. Momin, T. Khan, S. Gharat, R.S. Ningthoujam, A. Omri, Metallic nanoparticles as drug delivery system for the treatment of cancer, Expert Opin. Drug Deliv. 18 (2021) 1261–1290. https://doi.org/10.1080/17425247.2021.1912008

[35] P.K. Mishra, H. Mishra, A. Ekielski, S. Talegaonkar, B. Vaidya, Zinc oxide nanoparticles: a promising nanomaterial for biomedical applications, Drug Discov. Today. 22 (2017) 1825–1834. https://doi.org/10.1016/j.drudis.2017.08.006

[36] M.G. Vander Heiden, Targeting cancer metabolism: A therapeutic window opens, Nat. Rev. Drug Discov. 10 (2011) 671–684. https://doi.org/10.1038/nrd3504

[37] K. Seidi, H.A. Neubauer, R. Moriggl, R. Jahanban-Esfahlan, T. Javaheri, Tumor target amplification: Implications for nano drug delivery systems, J. Control. Release. 275 (2018) 142–161. https://doi.org/10.1016/j.jconrel.2018.02.020

[38] N. Hinge, M.M. Pandey, G. Singhvi, G. Gupta, M. Mehta, S. Satija, M. Gulati, H. Dureja, K. Dua, Nanomedicine advances in cancer therapy, Adv. 3D-Printed Syst. Nanosyst. Drug Deliv. Tissue Eng. (2020) 219–253. https://doi.org/10.1016/B978-0-12-818471-4.00008-X

[39] M. Xu, X. Han, H. Xiong, Y. Gao, B. Xu, G. Zhu, J. Li, Cancer Nanomedicine: Emerging Strategies and Therapeutic Potentials, Molecules. 28 (2023) 5145. https://doi.org/10.3390/molecules28135145

[40] J. Liu, J.F. Yan, Z.S. Deng, Nano-cryosurgery: A basic way to enhance freezing treatment of tumor, ASME Int. Mech. Eng. Congr. Expo. Proc. 2 (2007) 87–94. https://doi.org/10.1115/IMECE2007-43916

[41] Y. Hou, Z. Sun, W. Rao, J. Liu, Nanoparticle-mediated cryosurgery for tumor therapy, Nanomedicine Nanotechnology, Biol. Med. 14 (2018) 493–506. https://doi.org/10.1016/j.nano.2017.11.018

[42] A. Aderem, D.M. Underhill, Mechanisms of phagocytosis in macrophages, Annu. Rev. Immunol. 17 (1999) 593–623. https://doi.org/10.1146/annurev.immunol.17.1.593

[43] S. Xiang, H. Tong, Q. Shi, J.C. Fernandes, T. Jin, K. Dai, X. Zhang, Uptake mechanisms of non-viral gene delivery, J. Control. Release. 158 (2012) 371–378. https://doi.org/10.1016/j.jconrel.2011.09.093

[44] T.J. Pucadyil, S.L. Schmid, Conserved functions of membrane active GTPases in coated vesicle formation, Science (80-.). 325 (2009) 1217–1220. https://doi.org/10.1126/science.1171004

[45] A. Benmerah, C. Lamaze, Clathrin-coated pits: Vive la différence?, Traffic. 8 (2007) 970–982. https://doi.org/10.1111/j.1600-0854.2007.00585.x

[46] L.K. Medina-Kauwe, "Alternative" endocytic mechanisms exploited by pathogens: New avenues for therapeutic delivery?, Adv. Drug Deliv. Rev. 59 (2007) 798–809. https://doi.org/10.1016/j.addr.2007.06.009

[47] R.G. Parton, K. Simons, The multiple faces of caveolae, Nat. Rev. Mol. Cell Biol. 8 (2007) 185–194. https://doi.org/10.1038/nrm2122

[48] I.A. Khalil, K. Kogure, H. Akita, H. Harashima, Uptake pathways and subsequent intracellular trafficking in nonviral gene delivery, Pharmacol. Rev. 58 (2006) 32–45. https://doi.org/10.1124/pr.58.1.8

[49] L. Pelkmans, D. Püntener, A. Helenius, Local actin polymerization and dynamin recruitment in SV40-induced internalization of caveolae, Science (80-.). 296 (2002) 535–539. https://doi.org/10.1126/science.1069784

[50] J. Mercer, A. Helenius, Virus entry by macropinocytosis, Nat. Cell Biol. 11 (2009) 510–520. https://doi.org/10.1038/ncb0509-510

[51] S. Muzammil, J. Neves Cruz, R. Mumtaz, I. Rasul, S. Hayat, M.A. Khan, A.M. Khan, M.U. Ijaz, R.R. Lima, M. Zubair, Effects of Drying Temperature and Solvents on In Vitro Diabetic Wound Healing Potential of Moringa oleifera Leaf Extracts, Molecules. 28 (2023). https://doi.org/10.3390/molecules28020710

[52] N. Shadmani, K.H. Kahkesh, Fabrication of Biomimetic Cell Membrane-Functionalized Nanosystems, ACS Symp. Ser. 1464 (2024) 31–56. https://doi.org/10.1021/bk-2024-1464.ch003

[53] M. Liu, R.C. Anderson, X. Lan, P.S. Conti, K. Chen, Recent advances in the development of nanoparticles for multimodality imaging and therapy of cancer, Med. Res. Rev. 40 (2020) 909–930. https://doi.org/10.1002/med.21642

[54] M.R. Sheen, P.H. Lizotte, S. Toraya-Brown, S. Fiering, Stimulating antitumor immunity with nanoparticles, Wiley Interdiscip. Rev. Nanomedicine Nanobiotechnology. 6 (2014) 496–505. https://doi.org/10.1002/wnan.1274

[55] N. Naseri, E. Ajorlou, F. Asghari, Y. Pilehvar-Soltanahmadi, An update on nanoparticle-based contrast agents in medical imaging, Artif. Cells, Nanomedicine Biotechnol. 46 (2018) 1111–1121. https://doi.org/10.1080/21691401.2017.1379014

[56] S. Gavas, S. Quazi, T.M. Karpiński, Nanoparticles for Cancer Therapy: Current Progress and Challenges, Nanoscale Res. Lett. 16 (2021). https://doi.org/10.1186/s11671-021-03628-6

[57] M.R. Mohammadi, C. Corbo, R. Molinaro, J.R.T. Lakey, Biohybrid Nanoparticles to Negotiate with Biological Barriers, Small. 15 (2019). https://doi.org/10.1002/smll.201902333

[58] K.P. Das, J. Chandra, Nanoparticles and convergence of artificial intelligence for targeted drug delivery for cancer therapy: Current progress and challenges, Front. Med. Technol. 4 (2022). https://doi.org/10.3389/fmedt.2022.1067144

[59] B. Kumari, A. Hora, M.A. Mallick, Nanomedicines in Cancer Research: An Overview, LS Int. J. Life Sci. 6 (2017) 11. https://doi.org/10.5958/2319-1198.2017.00002.1

[60] S. Karaosmanoglu, M. Zhou, B. Shi, X. Zhang, G.R. Williams, X. Chen, Carrier-free nanodrugs for safe and effective cancer treatment, J. Control. Release. 329 (2021) 805–832. https://doi.org/10.1016/j.jconrel.2020.10.014

[61] J. Iqbal, B.A. Abbasi, R. Ahmad, T. Mahmood, B. Ali, A.T. Khalil, S. Kanwal, S.A. Shah, M.M. Alam, H. Badshah, A. Munir, Nanomedicines for developing cancer nanotherapeutics: from benchtop to bedside and beyond, Appl. Microbiol. Biotechnol. 102 (2018) 9449–9470. https://doi.org/10.1007/s00253-018-9352-3

Nanomaterials in Biological Systems
Materials Research Foundations 185 (2026) 112-153

Materials Research Forum LLC
https://doi.org/10.21741/9781644903858-6

Chapter 6

Organic Nanoparticle for Anti-Inflammatory Studies

Smritilekha Bera* and Dhananjoy Mondal

School of Chemical Sciences, Central University of Gujarat, Gandhinagar, Gujarat, India

lekha026@yahoo.com

Abstract

Nanobiotechnology plays a pivotal role in revolutionizing drug delivery and opens up exciting new possibilities for clinical treatment. The distinctive characteristics of nanoparticles make them a promising choice for drug delivery systems, particularly in the realm of anti-inflammatory therapies. Inflammatory conditions like inflammatory bowel disease, rheumatoid arthritis, and osteoarthritis are significantly impacted by inflammation. Utilizing nanoparticles for the efficient delivery of anti-inflammatory drugs can lead to reduced dosages and improved therapeutic outcomes. This review explores the potential of organic nanoparticles to exhibit anti-inflammatory properties and the prospects of nanomedicine in treating inflammatory diseases. In different pharmaceutical domains, applications of inorganic nanoparticles have already been well-versed. However, due to their toxicity and limited bioavailability, the focus has shifted towards exploring organic nanoparticles, which offer various advantages in respect of improved biocompatibility, biodegradability, ease of modification, and controlled and targeted drug release compared to inorganic nanoparticles. The inflammatory response is a critical factor in various ailments, and anti-inflammatory drugs aim to balance the immune and inflammatory responses. Recent research is focused on hybrid materials with anti-inflammatory effects, aiming for effective drug delivery. Conditions characterized by excessive production of inflammatory mediators, such as inflammatory bowel disease, rheumatoid arthritis, osteoarthritis, wound healing, and sepsis, may significantly benefit from various types of nanoparticles that have demonstrated anti-inflammatory potential.

Keywords

Nanomaterials, Drugs, Characterization, Anti-inflamatory

Contents

1.   Introduction

Inflammation is a pathological state that arises as a component of our immune system's reaction to detrimental triggers, encompassing irritants, damaged cells, and pathogens [1]. This complex biological reaction involves various cellular and molecular components working together to protect and heal the affected area. Importantly, inflammation assumes a substantial role in the progression of numerous illnesses [2]. The term "inflammare" in Latin means "To set on fire," which reflects the fiery nature of this immune response. While the innate immune response serves to protect the body against harmful agents, it also activates the innate inflammatory response system. Inflammation acts as a vital response to possible alarm signals and damage to organs within our bodies. Nevertheless, specific conditions like rheumatoid arthritis, Crohn's disease, ulcerative colitis, lupus, and similar ailments entail the immune system launching attacks on the body's own organs. These conditions can cause considerable pain, progressive debilitation, and significantly impact people's quality of life, as well as pose societal and economic burdens [3,4]. The objective of anti-inflammatory drugs is to restore the balance between inflammatory and immune responses. By regulating this interplay, these drugs aim to alleviate the symptoms and effects of inflammation-related diseases, offering relief and improving patients' well-being.

1.1   History of Inflammation

Around 3000 BC, an Egyptian papyrus referred to the four cardinal signs of inflammation, which were later elaborated on by Celsus in the 1st century AD. Subsequently, Virchow introduced the fifth clinical sign, known as "functio laesa". Interestingly, in 1973, John Hunter proposed that inflammation is not a disease but rather a non-specific response that actually benefits the host.

1.2   Anti-Inflammatory Diseases

The process of inflammation assumes a crucial role in regulating and facilitating the recovery of injuries within the body. It can be classified into two main types: acute inflammation and chronic inflammation.

i. Acute inflammation

Acute inflammation is an immediate response that occurs due to physical trauma, infection, stress, or a combination of these factors. It constitutes an essential aspect of the body's immune response, contributing to the prevention of additional harm while fostering the process of healing and recuperation. However, if inflammation persists and becomes self-sustaining, it can transform into chronic inflammation. Unlike acute inflammation, chronic inflammation extends beyond the initial injury and can last for extended periods, even months or years. This prolonged state can give rise to various health problems and may necessitate medical intervention to manage or prevent further damage caused by the persistent inflammation.

ii. Chronic inflammation

Chronic inflammation has the potential to impact various regions of the body and can additionally manifest as a secondary element in numerous illnesses. For example, conditions like atherosclerosis, where chronic inflammation affects the walls of blood vessels, accumulation of plaque, contribute to the blockages in arteries, and the onset of heart disease. Chronic inflammation also assumes a substantial role in other circumstances, including chronic pain, compromised obesity, sleep patterns, limitations in physical function, and a decline in overall well-being [5].

Inflammation can be classified into bacterial and non-bacterial categories, and its management typically entails the utilization of steroidal and non-steroidal anti-inflammatory drugs. In instances of bacterial inflammation, antibiotics are frequently administered in conjunction with these treatments. Chronic inflammation has the potential to give rise to systemic disorders, while acute inflammation can intensify and pose a risk to life. Various chronic diseases, including rheumatoid arthritis [6], atherosclerosis [7], diabetes [8], Alzheimer's [9], inflammatory bowel disease [10], and cancer [11], have been linked to inflammation. During inflammation, multiple cell types such as neutrophils, basophils, eosinophils, monocytes, and macrophages are involved [12-14]. Pro- and anti-inflammatory mediators, including cytokines, chemokines, tumor necrosis factor-α, cyclooxygenase-2 (COX-2), and inducible NO synthase (iNOS) play crucial regulatory roles [15].

Marine organisms, particularly soft corals found in coral reefs, offer a rich source of natural compounds with antimicrobial, antimalarial, antiviral, antioxidant, antitumor, and anti-inflammatory properties [16]. Soft corals contain secondary metabolites like sesquiterpenes [17], diterpenes [18], and sterols [19], which have demonstrated activities against fouling, viruses, tumors, and inflammation [20]. Notably, soft corals have yielded potent anti-inflammatory molecules such as isoparalemnone from the Formosan soft coral Paralemnalia thyrsoides [21].

However, traditional treatments for inflammation have limitations, including systemic adverse effects, unstable administration, gastrointestinal ulcer formation, insufficient local drug concentration, and challenges in maintaining drug levels. Despite the existing array of anti-inflammatory agents like glucocorticoids, nonsteroidal anti-inflammatory drugs, and immunosuppressants, their effectiveness is often suboptimal, and they are associated with undesirable side effects [22-24]. This emphasizes the importance of exploring novel candidates for anti-inflammatory drugs. Nanomedicine, employing nanoparticles, has arisen as an interdisciplinary domain that harnesses the physical, chemical, and biological attributes of nanoparticles. It has exhibited potential in the realm of anti-inflammatory therapy.

Nanomaterials in Biological Systems   Materials Research Forum LLC
Materials Research Foundations 185 (2026) 112-153   https://doi.org/10.21741/9781644903858-6

1.3   Fundamentals of Inflammation

Julius Cohnheim (1839-1884) is credited with describing the process of inflammation, while Ellie Metchnikoff is known for his discovery of phagocytosis. Additionally, Sir Thomas Lewis contributed to our understanding of inflammation by highlighting the role of chemical substances, such as locally induced histamine, in mediating the vascular changes associated with inflammation.

The concept of inflammation dates back thousands of years, with early observations made by ancient civilizations. However, it was in the 19th century that significant progress was made in understanding the basic mechanisms of inflammation. In the mid-1800s, German pathologist Rudolf Virchow described the four cardinal signs of inflammation: redness (rubor), swelling (tumor), heat (calor), and pain (dolor). These observations formed the basis for recognizing and diagnosing inflammation. Later, in the late 19th and early 20th centuries, the Russian scientist Elie Metchnikoff made groundbreaking discoveries related to inflammation. Metchnikoff focused on the role of immune cells, particularly phagocytes, in the inflammatory response. He observed that these cells engulf and destroy harmful microorganisms and debris, playing a crucial role in the body's defense against infection and injury. This process, known as phagocytosis, became a central concept in understanding inflammation.

Throughout the 20th century, research continued to unravel the complex mechanisms underlying inflammation. Scientists discovered various chemical mediators involved in the process, including histamine, prostaglandins, and cytokines. These mediators act as signaling molecules, orchestrating the inflammatory response and influencing the behavior of immune cells and blood vessels. Advancements in molecular biology and immunology further expanded our understanding of inflammation. Researchers identified specific receptors, such as Toll-like receptors, that recognize infectious agents and trigger inflammatory pathways. They also uncovered the intricate interactions between different immune cells, including neutrophils, macrophages, and lymphocytes, in orchestrating the immune response.

Today, the understanding of inflammation has evolved into a multidisciplinary field, encompassing various branches of biology and medicine. Inflammation is a double-edged sword, as it serves as a necessary defense mechanism and aids in tissue repair. Nevertheless, when inflammation becomes dysregulated or persistent, it can contribute to the onset of diverse ailments, encompassing autoimmune disorders, cardiovascular conditions, and cancer. Ongoing research is persistently revealing fresh perspectives on the intricacies of inflammation, thereby laying the groundwork for the creation of innovative therapeutic strategies aimed at regulating and managing the inflammatory reaction. Inflammation is a safeguarding reaction that engages host cells, blood vessels, proteins, and additional agents, with the goal of eradicating the initial source of cell damage, along with the necrotic cells and tissues arising from the primary injury. Its objective is to initiate the reparative process.

The basic mechanism of inflammation involves a complex series of events orchestrated by the immune system to protect the body from injury, infection, or tissue damage (Figure 1). The process can be divided into several key steps:

a. *Initiation*: Inflammation is typically triggered by a harmful stimulus, such as pathogens, physical trauma, or chemical irritants. Recognition of these stimuli by immune cells, particularly pattern recognition receptors (PRRs), initiates the inflammatory response.

Materials Research Forum LLC
https://doi.org/10.21741/9781644903858-6

b. *Vasodilation and increased vascular permeability*: In response to the stimulus, blood vessels near the affected area dilate, allowing increased blood flow. This causes redness and heat, two characteristic signs of inflammation. Increased permeability of the blood vessel walls also occurs, leading to the leakage of fluid and immune cells into the surrounding tissue.

c. *Recruitment of immune cells*: Chemical signals, including cytokines and chemokines, are released at the site of inflammation. These signals attract immune cells, such as neutrophils and macrophages, to the affected area. Neutrophils are the first responders, arriving quickly but having a short lifespan. Macrophages arrive later and play a key role in phagocytosis and initiating tissue repair.

d. *Phagocytosis and destruction of pathogens*: Immune cells, notably neutrophils and macrophages, undertake the task of engulfing and neutralizing pathogens via a mechanism known as phagocytosis. This helps to eliminate the source of the inflammation and prevent further infection.

e. *Release of inflammatory mediators*: Immune cells and damaged tissues release various chemical mediators, including histamine, prostaglandins, and cytokines. These mediators contribute to the inflammatory response by causing further dilation of blood vessels, attracting more immune cells, and promoting the activation of other cells involved in the immune response.

f. *Tissue repair and resolution*: Once the harmful stimulus is neutralized and the pathogens are eliminated, the inflammatory response enters a resolution phase. During this phase, anti-inflammatory mediators are released, and tissue repair processes begin. The immune cells involved in the inflammatory response shift their function towards promoting tissue healing and resolution of inflammation.

It is essential to emphasize that although inflammation is a critical defense mechanism, it requires tight regulation. Imbalanced or prolonged inflammation can result in tissue harm and play a role in the emergence of a range of disorders, including autoimmune conditions, cardiovascular ailments, and persistent inflammatory diseases.

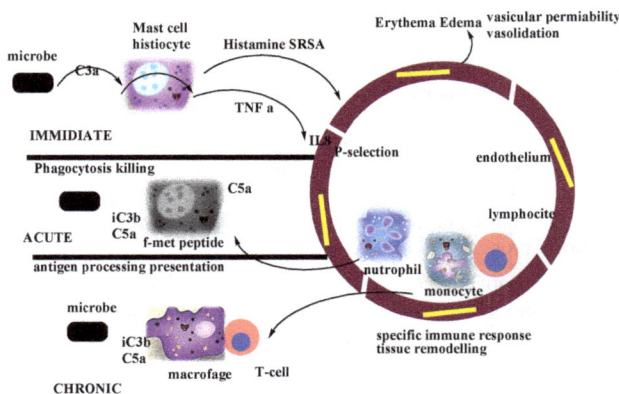

Figure 1. Basic mechanism of inflammation

1.4 Common Types of Drugs for Anti-Inflammatory Response

Anti-inflammatory drugs are used to decrease inflammation, relieve pain, and alleviate symptoms associated with inflammatory conditions. There are several common types of drugs used to elicit an anti-inflammatory response in various medical conditions.

Nonsteroidal anti-inflammatory drugs (NSAIDs): NSAIDs are a commonly used class of anti-inflammatory drugs. These drugs function by inhibiting the activity of enzymes called cyclooxygenases (COX), which are responsible for generating prostaglandins that facilitate inflammation and pain. Notable examples of NSAIDs encompass naproxen, ibuprofen, aspirin, and diclofenac (Figure 2).

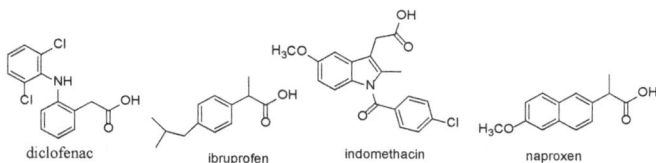

Figure 2. A few NSAID drugs for inflammation

Steroidal anti-inflammatory drugs: Steroidal anti-inflammatory drugs, often referred to as corticosteroids or glucocorticoids, imitate the actions of hormones naturally produced by the adrenal glands. Possessing strong anti-inflammatory attributes, they find utility in a variety of inflammatory disorders. These medications can be applied topically, taken orally, or administered through injections, depending on the intensity and site of inflammation. Notable examples encompass dexamethasone, prednisone, and hydrocortisone (Figure 3).

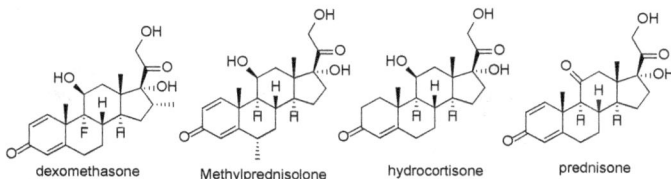

Figure 3. A few steroidal anti-inflammatory drugs

Disease-modifying anti-rheumatic drugs (DMARDs): DMARDs (Disease-Modifying Antirheumatic Drugs) constitute a collection of medications employed to manage persistent inflammatory conditions like psoriatic arthritis, rheumatoid arthritis, and systemic lupus erythematosus. These drugs operate by dampening the immune system's activity and mitigating inflammation. Illustrative instances involve methotrexate, sulfasalazine, hydroxychloroquine, along with biologic DMARDs such as tumor necrosis factor (TNF) inhibitors and interleukin-6 (IL-6) inhibitors (Figure 4).

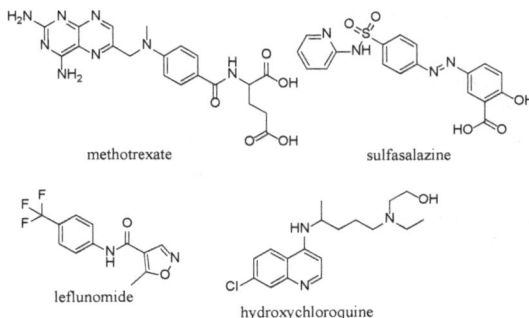

Figure 4. A few disease-modifying anti-rheumatic drugs

d. *Biologic therapies*: Biologic therapies are a newer class of anti-inflammatory drugs that target specific molecules or cells involved in the inflammatory process. They are often used when other treatments have been ineffective or are not well-tolerated. Biologics include TNF inhibitors, IL-6 inhibitors, interleukin-1 (IL-1) inhibitors, and monoclonal antibodies targeting other immune pathways. These drugs are typically administered by injection or infusion.

e. *Topical anti-inflammatory drugs*: Topical anti-inflammatory drugs are applied directly to the skin or affected area to reduce inflammation and relieve localized pain. Examples include creams, gels, ointments, and patches containing NSAIDs or other anti-inflammatory agents.

f. *Over-the-counter (OTC) anti-inflammatory drugs*: Some mild to moderate anti-inflammatory drugs, such as certain NSAIDs, are available over-the-counter without a prescription. These medications are generally used for temporary relief of pain and inflammation associated with conditions like headaches, minor injuries, and menstrual cramps. However, it is important to follow the recommended dosage and consult a healthcare professional if symptoms persist or worsen.

It is crucial to note that while anti-inflammatory drugs can effectively manage inflammation and provide relief, they may have potential side effects and interactions with other drugs. Prior to initiating any new medication, it is recommended to seek guidance from a healthcare provider, and to adhere to the prescribed dosage and duration for the use of these drugs.

1.5   Nanoparticles in Drug Development

The prominence of nanoparticle-based technology has increased with the advancement of a new generation of promising anti-inflammatory nanoparticle agents. These nanoparticles possess exceptional physicochemical and functionalization properties that help overcome the limitations of traditional anti-inflammation methods. The application of nanotechnology in drug delivery techniques has the potential to overcome numerous challenges faced by conventional methods [25]. This is mainly due to the unique and well-defined physicochemical properties exhibited by nanoparticles, such as ultra-small and controllable quantum-size effects, specific shape characteristics, high surface-to-volume ratio [26], functionalizable structure, surface charge, and distinct electrodynamical interactions [27-30]. In various pharmaceutical fields, inorganic applications have already been developed. However, their toxicity and limited bioavailability have led to the exploration of organic nanoparticles, which are biocompatible, biodegradable, easily modifiable, and enable controlled and targeted drug release. Organic or organic-inorganic hybrid nanoparticles offer significant advantages as they possess properties like biodegradability, biocompatibility, bioavailability, and ease of modification [31]. Several nanoparticle-based strategies have been reported, including the formation of bio-polymers, liposomes, dendrimers, niosomes with carbohydrate coatings, and glycosidic frameworks [32-34]. These approaches not only present intriguing prospects for advancements in synthetic organic chemistry but also play a role in the advancement of biomaterials and novel techniques for drug delivery [35,36]. Therefore, there is a demand to develop novel therapeutics with reduced toxicity and increased bioavailability and biocompatibility, either through organic nanoparticles or the hybridization of organic and inorganic nanoparticles. Nanoparticles can be utilized for physical encapsulation, adsorption, or chemical conjugation in these endeavors.

2.   Nanoparticle-Based Drugs for Anti-Inflammatory Treatment

Nanoparticle-based drugs have shown great promise in the field of anti-inflammatory treatment. These innovative drug delivery systems utilize nanoparticles as carriers to enhance the therapeutic efficacy and targeted delivery of anti-inflammatory agents. Here is some information on nanoparticle-based drugs for anti-inflammatory treatment:

a. *Enhanced drug delivery*: Nanoparticles can encapsulate anti-inflammatory drugs, protecting them from degradation and improving their stability. This allows for controlled release and prolonged circulation of the drugs in the body, enhancing their bioavailability and therapeutic effect.

b. *Targeted delivery*: Nanoparticles can be designed to specifically target inflamed tissues or cells, minimizing off-target effects and reducing systemic toxicity. Surface modifications of nanoparticles, such as ligand conjugation or antibody targeting, enable selective interaction with inflamed areas, improving drug accumulation and efficacy at the site of inflammation.

c. *Anti-inflammatory drug incorporation*: Nanoparticles can be loaded with various anti-inflammatory agents, including nonsteroidal anti-inflammatory drugs (NSAIDs), glucocorticoids, immunosuppressants, and biologics such as anti-inflammatory peptides or antibodies. These drugs can modulate inflammatory pathways, inhibit immune responses, or target specific molecular targets involved in inflammation.

d. *Sustained release*: Nanoparticles provide sustained release profiles, allowing for prolonged therapeutic action of anti-inflammatory drugs. This sustained release can reduce the frequency of drug administration, improve patient compliance, and provide a more stable therapeutic effect over time.

e. *Localized therapy*: Nanoparticles can be administered directly at the site of inflammation, such as joints in the case of arthritis or diseased tissues. This localized therapy minimizes systemic exposure, reduces side effects, and maximizes drug concentration at the site of action.

f. *Combination therapy*: Nanoparticles offer the possibility of co-loading multiple anti-inflammatory drugs or combining anti-inflammatory agents with other therapeutic agents such as analgesics, antioxidants, or regenerative factors. This combination therapy approach can provide synergistic effects and address multiple aspects of inflammation simultaneously.

Materials Research Forum LLC
https://doi.org/10.21741/9781644903858-6

g. *Imaging and diagnostic capabilities*: Some nanoparticles possess inherent imaging properties or can be functionalized with imaging agents. This enables real-time monitoring of inflammation, assessment of treatment efficacy, and personalized therapy.

Nanoparticle-based drugs for anti-inflammatory treatment represent a rapidly advancing field with significant potential for improving therapeutic outcomes, reducing side effects, and revolutionizing the management of inflammatory conditions. Ongoing research and development in this area hold great promise for the future of anti-inflammatory therapy.

3. Nanoparticles for Anti-Inflammatory Drug Development

Organic nanoparticle for inflammatory drug development: Organic nanoparticles (NPs) are known for their improved biocompatibility compared to inorganic nanoparticles. They are widely used as carriers for various applications, including drug delivery, imaging, and diagnostics. Carriers for organic NPs are typically composed of biomaterials or polymers. These materials serve as the matrix or shell that encapsulates or binds the active payload, such as drugs or imaging agents. These are just a few examples of natural and synthetic polymers used as carriers for organic nanoparticles. The choice of polymer depends on factors such as biocompatibility, stability, drug release kinetics, and the specific requirements of the application.

Natural polymers: These polymers are derived from natural sources such as proteins, polysaccharides, or nucleic acids. Examples of natural polymers commonly used in NP formulations include:

Chitosan: Derived from chitin, a natural polymer found in the exoskeleton of crustaceans. Chitosan-based NPs have been extensively studied for drug delivery and gene therapy applications.

Gelatin: Derived through partial hydrolysis of collagen, gelatin finds extensive use in biomedical contexts. Nanoparticles (NPs) crafted from gelatin exhibit remarkable biocompatibility and have been investigated for their potential in drug delivery and tissue engineering.

Albumin: A protein found in blood plasma, albumin-based NPs have shown promise for drug delivery due to their biocompatibility and ability to carry hydrophobic drugs.

Hyaluronic acid: A naturally occurring polysaccharide, hyaluronic acid-based NPs are commonly used for targeted drug delivery and tissue engineering applications.

Synthetic polymers: These polymers are chemically synthesized in the laboratory, offering precise control over their properties and structure. Some commonly used synthetic polymers for organic NP carriers include:

Poly(lactic-co-glycolic acid) (PLGA): A biodegradable and biocompatible polymer that has been extensively investigated for drug delivery applications. PLGA-based NPs can provide sustained drug release and are FDA-approved for certain applications.

Polyethylene glycol (PEG): Known for its "stealth" properties, PEGylation involves coating NPs with PEG, which helps to evade the immune system and prolong their circulation time in the body.

Poly(caprolactone) (PCL): Another biodegradable polymer often used in NP formulations. PCL-based NPs offer controlled release properties and have been explored for drug delivery and tissue engineering.

Poly(amidoamine) (PAMAM): Dendrimers are a class of highly branched polymers, and PAMAM dendrimers have been utilized as carriers for drug delivery and gene delivery due to their unique architecture and functionalization capabilities.

This review article discusses the use of a few polymers for their anti-inflammatory properties. Nowadays, inflammatory bowel disease, rheumatoid arthritis, and osteoarthritis are only a few of the diseases that might arise when there is excessive production of inflammatory mediators. Lin et al. developed reducible thermosensitive PEGylated poly(N-isopropylacrylamide) (pNIPAM) nanoparticles with degradable disulfide crosslinks, referred to as NGPEGSS [37]. These nanoparticles were designed to encapsulate and release anti-inflammatory MK2 inhibiting peptides with the sequence KAFAKLAARLYRKALARQLGVAA (abbreviated as KAFAK) (Figure 5). Through a controlled and sustained drug release mechanism, the nanoparticles efficiently delivered the anti-inflammatory peptides to the inflamed cartilage tissue via passive targeting.

In vitro experiments using cartilage cells demonstrated that the released anti-inflammatory peptides effectively suppressed the production of inflammatory molecules, leading to a reduction in cartilage inflammation. Moreover, when compared to non-degradable counterparts, the KAFAK-loaded NGPEGSS treatment exhibited a greater extent of reduction in *ex vivo* inflammation. This study highlights the potential of these reducible thermosensitive nanoparticles as a promising approach for targeted and sustained delivery of anti-inflammatory peptides, showing encouraging results in reducing cartilage inflammation.

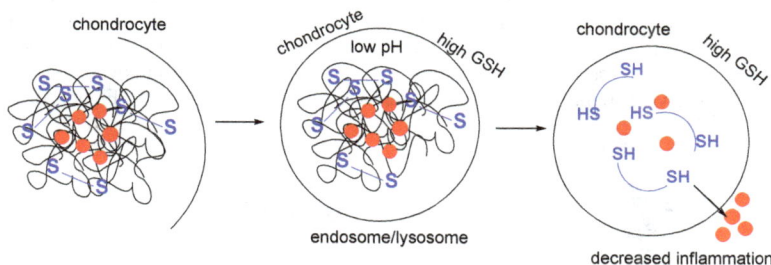

Figure 5. Pictorial diagram of thermosensitive PEGylated poly(N-isopropyl acrylamide) nanoparticles with degradable disulphide cross-links

Chang et al. designed liposomal nanoparticles containing dexamethasone (an anti-inflammatory drug) and moxifloxacin (an antibiotic) to enhance drug delivery and retention in the cornea, thereby improving treatment efficacy for wound healing [38]. To further augment the therapeutic effect and create a favorable environment for wound healing, the liposomal nanoparticles were embedded in a collagen/gelatin/alginate hydrogel. The resulting nanostructured lipid carriers loaded with moxifloxacin and dexamethasone (Lipo-MFX/DEX) were combined with the biocompatible hydrogel to form CGA-Lipo-MFX/DEX. This combination allowed for efficient drug release, achieving the desired drug concentration within just 60 minutes and sustaining drug release for at least 12 hours. The incorporation of the hydrogel provided support for the healing process and facilitated controlled drug release, contributing to the overall effectiveness of the treatment for wound healing. The study demonstrates the potential of CGA-Lipo-MFX/DEX as a promising strategy for improved drug delivery and therapeutic outcomes in corneal wound healing. Crucially, the formulated CGA-Lipo-MFX/DEX showed no significant toxic effects. When tested on ocular epithelial cells, it not only demonstrated biocompatibility but also exhibited the potential to stimulate cell growth, suggesting its ability to promote cell proliferation and tissue repair. These findings underscore the promising applications of CGA-Lipo-MFX/DEX in ocular therapies, offering targeted and sustained drug delivery with beneficial effects on cell growth and wound healing. Both *in vitro* and *in vivo* experiments supported the effectiveness of the liposomal nanoparticle-hydrogel combination in suppressing bacterial growth, reducing corneal inflammation, and facilitating corneal wound healing.

Several varieties of nanoparticles (NPs) have demonstrated promising anti-inflammatory attributes [39]. Inflammatory bowel diseases (IBD), such as Crohn's disease and ulcerative colitis, are enduring intestinal inflammatory conditions that frequently demand frequent or continuous administration of anti-inflammatory medications at high doses [40]. Nanoparticles engineered to regulate drug release precisely at the intended location can facilitate treatment and diminish potential side effects [41-43].

For instance, Lamprecht et al. developed Tacrolimus nanoparticles (NP) using biodegradable poly(lactide-co-glycolide) (PLGA) or pH-sensitive Eudragit P-4135F, which exhibited promising results in treating inflamed bowel tissue associated with inflammatory bowel disease (IBD). In a mouse model of colitis induced by dextran sulfate, every version of tacrolimus formulations exhibited a decrease in clinical activity score and myeloperoxidase activity, simultaneously leading to a notable increase in colon length. Subcutaneous drug solutions were more effective than oral NP formulations, but both NP approaches were superior to oral drug solutions. Additionally, both NP strategies exhibited potential in reducing nephrotoxicity, with the pH-sensitive NP showing slightly fewer adverse effects related to nephrotoxicity (Figure 6).

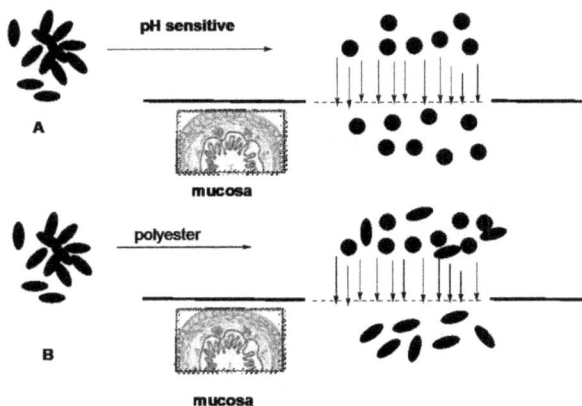

Figure 6. Application of NP in oral administration of anti-inflammatory drugs

Nanomaterials in Biological Systems   Materials Research Forum LLC
Materials Research Foundations 185 (2026) 112-153   https://doi.org/10.21741/9781644903858-6

Rolipram, an exemplar anti-inflammatory drug, was encapsulated within nanoparticles made of poly(lactic-co-glycolic acid) (PLGA). These nanoparticles were administered orally once daily for a continuous span of five days. All nanoparticle formulations exhibited efficacy equivalent to the drug in its soluble form in alleviating experimental colitis. Following oral administration of both rolipram nanoparticles and the solution, there was a substantial reduction in clinical activity score and myeloperoxidase activity. Notably, the group treated with rolipram solution demonstrated a high adverse effect index. In contrast, the groups receiving rolipram nanoparticles displayed their potential to restrict systemic drug absorption, as evidenced by a marked reduction in the adverse effect index.

Rheumatoid arthritis is a persistent autoimmune ailment marked by joint inflammation and damage. Glucocorticoids are recognized as a conventional and effective therapeutic approach for Rheumatoid arthritis. Nevertheless, Glucocorticoids treatment lacks specificity and imparts systemic effects. Extended use of these treatments results in severe undesirable effects. However, ongoing research concerning nanoparticles explores a drug delivery system that aims to target the delivery of medications, with the intention of diminishing systemic adverse reactions [44]. As a potential platform for treating rheumatoid arthritis, Shi et al. concentrated on cell membrane nanovesicles that express TRAIL (tumor necrosis factor-related apoptosis-inducing ligand). It was made using poly (DL-lactic-co-glycolic acid) to attach TRAIL-expressing HUVEC cell membranes to drug-loaded polymeric cores. Inducing apoptosis (cell death) in specific cell types, especially inflammatory cells, is one of the functions of the protein TRAIL. Fluorescence and photoacoustic dual-modal imaging revealed enhanced accumulation and prolonged retention of TU-NPs (Tumor Necrosis Factor-α-Targeted Nanoparticles) within inflamed joints following intravenous administration in animals with collagen-induced arthritis. These nanoparticles appeared to have substantial chondroprotection against joint damage, the ability to target and deeply penetrate inflamed tissues, and the ability to neutralize cytokines, decrease synovial inflammation, and neutralize other cytokines *in vivo* therapeutic evaluations (**Figure 7**).

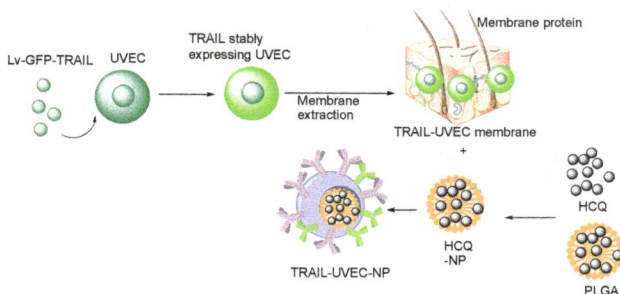

Figure 7. Preparation of TU-NPs for targeting RA

Furthermore, acute inflammatory conditions, including sepsis [45], trauma [46] and the sudden onset of chronic diseases, pose greater complexity than chronic inflammatory ailments. The management of these conditions demands swift and precise drug delivery. Nanoparticles, due to their diminutive size, exhibit the ability to concentrate drugs within the intended tissue, thereby enhancing the efficacy of disease treatment. In recent times, drug delivery systems based on nanoparticles have showcased the potential to heighten anti-inflammatory outcomes while diminishing adverse effects. Additionally, novel drug delivery systems have the capacity to amplify the efficacy of conventional medications.

3.1   Poly(Lactic-Co-Glycolic Acid) (Plga) Nanoparticles

Poly(lactic-co-glycolic acid) (PLGA) is a commonly employed polymer known for its biodegradability and compatibility with biological systems. It finds extensive use in crafting nanoparticles designed for drug delivery applications [47]. Due to its relatively hydrophobic nature, PLGA has emerged as a highly promising candidate for drug delivery, exhibiting significant potential in targeted therapy [48]. Incorporating drugs into PLGA and creating complexes is a extensively investigated strategy within drug delivery research.

Thevenot et al. established a mice implantation model, wherein stromal cell-derived factor-1α (SDF-1α) was integrated into PLGA [49]. This solid scaffold demonstrated effective reduction of the inflammatory response upon implantation in the subcutaneous region.

Fredman et al. conducted a study where they utilized an aminoterminal peptide comprising amino acids 2-26 (Ac2-26) with properties mimicking annexin A1 [50]. They encapsulated Ac2-26 with collagen IV (Col IV)-targeted PLGA to create Col IV-Ac2-26 PLGANPs. These nanoparticles demonstrated significant therapeutic efficacy in an atherosclerosis mice model. In the context of bone diseases, PLGA has found applications in conditions such as osteosarcomas, osteoarthritis, bone cancer metastasis, and other inflammatory bone diseases [51]. Feng et al. carried out research where they incorporated doxycycline (DOXY) into PLGA nanospheres, forming a 3D scaffold, and conducting in vitro tests [52]. Their findings revealed that DOXY-PLGA NPs effectively inhibited the growth of E. coli and S. aureus, and they also exhibited controlled release of DOXY.

Peng et al. introduced a biodegradable, thermosensitive implant comprised of poly(ethylene glycol) monomethyl ether (mPEG) and PLGA, which resulted in a solution-to-hydrogel (sol-gel) drug delivery system [53]. Additionally, PLGA was utilized to create a poly(ethylene glycol) monomethyl ether (mPEG) and PLGA copolymer, also serving as a sol-gel drug delivery system for treating osteomyelitis. These systems offer several advantages, including ease of preparation, 100% encapsulation rate for drugs, near-linear sustained drug release, injectable design, and in situ gelling at the target tissue. The implantation of the mPEG-PLGA hydrogel containing teicoplanin demonstrated effectiveness in treating osteomyelitis in a rabbit model, as supported by histological examination and immunoblotting analyses. The study highlights the potential of these biodegradable implants as promising drug delivery systems for osteomyelitis treatment with desirable sustained drug release and positive therapeutic outcomes (Scheme 1).

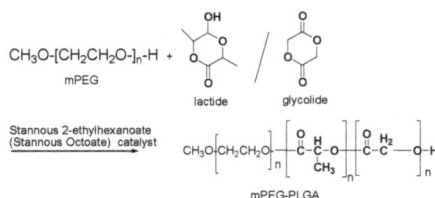

Scheme 1. Synthesis of mPEG-PLGA deblock copolymer

Trujillo-Nolasco et al. discovered that Lutetium-177 (177Lu) exhibited the capacity to diminish inflammation in synovial tissue [54]. Furthermore, they employed 1,4,7,10-Tetraazacyclododecane-1,4,7,10-tetraacetic acid (DOTA) as a component for 177Lu, which was synthesized utilizing hyaluronic acid (HA). The resulting complex, 177Lu-DOTA, was encapsulated alongside MTX (methotrexate) within PLGA. This construct, known as Lu-DOTA-HA-PLGA(MTX), displayed potential as a drug delivery system for anti-rheumatic therapy. However, further in vivo models and testing are required to validate these findings. Inflammatory bowel disease (IBD) constitutes a group of chronic gastrointestinal disorders, encompassing ulcerative colitis and Crohn's disease [55].

Although low molecular weight heparin (LMWH) is effective in ulcerative colitis, it lacks a specific drug delivery method. LMWH was combined with polymethacrylate nanoparticles (NP) of type A (PEMT-A) or type B (PEMT-B) with a size of 150 nm, yielding a maximum drug loading of 0.1 mg/mg. Both free LMWH and LMWH-NP dramatically reduced cytokine release in lipopolysaccharide-stimulated macrophages, independent of cellular absorption. The in-vivo therapeutic efficiency was dose-dependent, with only slight differences between free LMWH and LMWH-NP detected at low doses (100 IU/kg), and the superiority of LMWH-NP becoming apparent with a dose increase (500 IU/kg). When compared to LMWH, administration of LMWH-NP at 500 IU/kg significantly enhanced clinical activity, while pathophysiological indicators suggested an improved therapeutic outcome.

3.2   Chitosan Nanoparticles for Anti-Inflammatory Effect

Tahara et al. found that oral decoy oligonucleotide (ODN) was convenient for UC mouse models [56]. A method for oral delivery of a nuclear factor kappa B (NF-kB) decoy oligonucleotide (ODN) was tested in an experimental ulcerative colitis model. Chitosan (CS)-modified poly(D, L-lactide-co-glycolide) nanospheres (CS-PLGA NS) carrying the decoy ODN were created using an emulsion solvent diffusion technique. CS-PLGA NS demonstrated efficient cellular uptake compared to plain-PLGA NS in Caco-2 cells. The CS-PLGA NS protected the decoy ODN from degradation by DNase I and acidic media, such as gastric juice. Daily oral administration of CS-PLGA NS to rats significantly reduced ulcerative colitis symptoms, including myeloperoxidase activity, bloody stools, diarrhoea, and colon lengthening caused by dextran sulfate sodium. Moreover, the decoy ODN-loaded CS-PLGA NS specifically targeted and adhered to the inflamed mucosal tissue of ulcerative colitis model rats.

Furthermore, Zhang et al. developed a CS vector for delivering gene therapy, and DNA sequences encoding IL-1Ra and IL-10 were sent, and it was discovered that they were successful in preventing the onset of osteoarthritis in a rabbit model [57].

In order to improve theophylline's anti-inflammatory properties when administered intranasally, Lee et al. produced thiolated chitosan nanoparticles (TCNs) [58]. They employed BALB/c mice with allergic asthma, ovalbumin sensitizing them and creating an allergic inflammatory state. Theophylline was given to the mice either alone, in combination with chitosan nanoparticles, or with theophylline adsorbed on TCNs. When used alone, theophylline had mild anti-inflammatory benefits, including a decrease in eosinophils in BAL fluid, a reduction in bronchial damage, inhibition of mucus hypersecretion, and an increase in lung cell death. Theophylline's effects were markedly improved by the addition of thiolated chitosan nanoparticles.

In a similar manner, MAP4K4 siRNA was coupled with galactosylated trimethyl chitosan cysteine nanoparticles and given orally to treat ulcerative colitis by Zhang et al. It was discovered that the colon's MAP4K4 and TNF-mRNA levels, as well as TNF protein expression and MPO activity, were decreased by this treatment. Targeting active macrophages is advantageous, as demonstrated by the improved efficacy of galactosylated particles over non-galactosylated nanoparticles [59].

Melatonin-loaded chitosan nanoparticles (Mel-CSNPs) for inflammatory bowel disease to enhance drug release profile and assess its in-vitro and in-vivo therapeutic efficacy, Soni et al. In an in-vitro and in-vivo IBD model, Mel-CSNPs had a stronger anti-inflammatory effect. Nitric oxide (NO) reduction, suppressed nuclear translocation of NF-kB p65, and decreased IL-1 and IL-6 expression are all credited with Mel-CSNPs' significant anti-inflammatory activity [60].

Pertuit et al. synthesised nanoparticles with 5-amino salicylic acid for the treatment of inflammatory bowel disease [61,62]. 5-amino salicylic acid (5ASA) was covalently bonded to poly(caprolactone), and the nanoparticle matrix was obtained by oil/water emulsification or nanoprecipitation techniques. Toxicity of the various formulations was assessed using Caco-2 and HEK cell cultures, and it was shown to be marginally higher for 5ASA grafted NP compared to blank NP (Me5ASA-NP: 75 g/l; blank NP: 210 g/l). In vivo, all 5ASA-containing formulations reduced clinical activity score and myeloperoxidase activity (untreated control: 28.0 5.6 U/mg; 5ASA-NP (0.5 mg/kg): 15.2 5.6 U/mg; 5ASA solution (30 mg/kg): 16.2 3.6 U/mg). The use of NP formulations allowed for a considerable reduction in the dose of 5ASA.

Tang et al. reported high loading efficiency in a sustained release method for mesalazine (MSZ) by making hydroxypropyl-β-cyclodextrin (HP-b-CD) inclusion complex loaded chitosan (CS) nanoparticles (NPs) (HP-b-CD/MSZ/CS NPs) in a 1:1 ratio. When compared to free MSZ, the HP-CD/MSZ/CS NPs displayed a sustained release of MSZ, according to Tang et al. By keeping an eye on important inflammatory mediators, the activity of the NPs against a cytokine-triggered inflammatory response was assessed in cytokine-stimulated HT-29 cell lines. The findings showed that the NPs more potently suppressed the generation of NO, PGE2, and IL-8 when compared to free MSZ, suggesting that the NPs may have had superior anti-inflammatory effects [63].

Barbosa et al. created fucoidan/chitosan nanoparticles to deliver methotrexate topically for the treatment of skin-related inflammatory disorders. Pro-inflammatory cytokines generated by activated human monocytes are significantly decreased by MTX-loaded fucoidan/chitosan nanoparticles. Studies on skin permeation revealed that methotrexate-loaded nanoparticles were able to cross the pig ear skin barrier after 6 hours, which is a 2.7- and 3.3-fold improvement over free methotrexate [64].

By using the spray drying process, Oztürk et al. produced chitosan nanoparticles (CS-NPs) loaded with dexketoprofen trometamol (DT) for oral medication delivery. In both release media, it was found that the Weibull kinetic model best predicted DT release from CS-NPs. The findings of the anti-inflammatory activity tests showed that the M-DT coded NPs formulation had a significant amount of anti-inflammatory potential and a near inhibitory value to the conventional anti-inflammatory DT at a dosage that was one-fifth less [65].

Tetracycline-resistant enterotoxigenic E. coli (ETEC) was quickly destroyed by CNMs during various growth phases, and Tet-resistant ETEC was significantly eliminated from food-related biological settings. CNMs dramatically decreased the cytotoxicity of LPS-induced RAW264.7 macrophage cytotoxicity. By lowering the levels of nitric oxide and proinflammatory cytokines like tumor necrosis factor-, interleukin (IL)-6, IL-8, IL-1, Toll-like receptor 4, and nuclear factor-B (NF-B), in LPS-challenged RAW264.7 macrophages, CNMs significantly reduced the inflammatory response caused by LPS. The NF-B and mitogen-activated protein kinase signaling pathways were downregulated by CNMs, which prevented inflammatory reactions to LPS stimulation [66].

By using an ionotropic gelation process, Hirva et al. synthesized chitosan-alginate nanoparticles that were loaded with the NSAID (non-steroidal anti-inflammatory medication) zaltoprofen. Zaltoprofen's anti-inflammatory effects are caused by its ability to prevent bradykinin from interacting with the bradykinin 2 receptor at nerve terminals, which is followed by a suppression of bradykinin-induced reactions. The scientists created a zaltoprofen formulation with a prolonged release that has fewer adverse effects, greater bioavailability, and better patient compliance. On a carrageenan-induced paw-edema model in Wistar rats, the chitosan-alginate zaltoprofen gel's anti-inflammatory effectiveness was evaluated. In contrast to the conventional zaltoprofen gel, their findings demonstrated that the gel loaded with zaltoprofen nanoparticles displayed anti-inflammatory efficacy over a sustained period of time (10 h) [67].

Hazliza et al. looked into a percutaneous medication delivery system using chitosan nanoparticles (CS NPs) loaded with hydrocortisone (HC). The treatment was intended for atopic dermatitis, a disorder marked by chronic inflammation, extreme itching, and dry, scaly skin. In this study, CS NPs were made using the ionic gelation process, and HC was loaded into the CS using TPP. Commercial QV cream significantly increased the release and penetration of HC, whereas CS NPs offer a more consistent release. As a result, it can be regarded as a viable option for the delivery of the anti-inflammatory medication HC with increased effectiveness and decreased toxicities [68].

Mehmet et al. created a chitosan nanoparticle filled with ketorolac tromethamine for better ketorolac delivery in ocular disorders. The developed system showed a burst release at first, then a steady flow for up to 72 hours. The sustained release of ketorolac from the system indicates the suitability of the devised nano-therapeutic system for ocular inflammatory disorders [69].

Shah et al. created a skin-penetrating nanogel (SPN) for the topical administration of the anti-inflammatory medicines spantide II (SP) and ketoprofen (KP). The development of the nanoparticle system (NPS) involved the use of biodegradable, non-toxic polymers PLGA as the inner layer with SP and chitosan as the outer layer with KP. Oleic acid (NPSO) was used to modify the surface of the bi-layered system in order to improve penetration. Finally, to increase the skin contact time (SPN), the created NPSO was added to a gel matrix made of carbopol and hydroxypropyl methyl cellulose. Ex-vivo skin permeation testing on cadaveric human skin revealed improved SPN gel penetration and deposition. In comparison to SP-gel, SPN gel maintained 1.5 times and 2.7 times more SP in the epidermis and dermis, respectively. It was 1.4 and 2.1 times greater in the case of KP than the matching KP-gel. According to research on the anti-inflammatory effects of SPN, using it for three straight days dramatically decreased the ear thickness of C57BL/6 mice with di-fluoro nitrobenzene-induced allergic contact dermatitis from 148.46 to 56.23 mm [70].

When encapsulated in nanoparticles, Ac2-26, a peptide resembling annexin-A1, demonstrated enhanced efficacy as reported by Kamaly et al. Using the nanoprecipitation method, a collagen IV-targeting conjugate was employed to generate nanoparticles from a biodegradable PLGA-β-PEG polymer. These Ac2-26-loaded targeting nanoparticles outperformed non-targeted, jumbled peptide-loaded, and non-loaded particles in their ability to reduce the number of macrophages in an in vivo murine peritonitis model. In an *in vivo* ischemia model, targeted particles performed non-targeting peptide-loaded and modified peptide-loaded particles, with myeloperoxidase activity reduced [71].

Gadde et al. used a similar formulation to deliver an anti-inflammatory synthetic liver X receptor agonist (GW3965) as nanoparticles, but without the collagen IV targeting component. In vitro, treatment with GW3965 nanoparticles reduced monocyte chemoattractant protein 1 (MCP-1) and TNF-expression in peritoneal macrophages. Fewer macrophages were observed in an in vivo model of peritonitis, and MCP-1 and TNF-α-gene and protein expression were lowered to levels comparable to dexamethasone therapy [72].

A different study generated chitosan nanoparticles with a spacer arm length created by altering the number of glycine units in an arginine-containing peptide for the delivery of siRNA for the treatment of pulmonary inflammations. Studies on human alveolar epithelial cell lines performed in vitro and tests on mice used in a model of pulmonary inflammation performed in vivo showed that nanoparticles with the spacer arm length were superior to unmodified chitosan nanoparticles for the transfer of genes. When compared to unmodified chitosan, the association of the nanoparticle system with T cells was much stronger and the expression of CD4 receptors on T cells was significantly lower when examined using Jurkat and A301 cell lines. Compared to other carriers, chitosan nanoparticles had a better level of siRNA stability. When compared to naked siRNA, which degraded significantly more quickly in serum, the stability of siRNA loaded in chitosan nanoparticles was higher. This nanoparticle may be effective in treating inflammatory conditions including psoriasis and RA [73].

To combat the allergic and inflammatory reactions seen in asthma, chitosan nanoparticles loaded with plasmid DNA expressing for the IFN-α-gene were produced. Ovalbumine-induced inflammation model in mice was studied in vivo, and the results showed reduced airway hyperresponsiveness and inflammation as well as decreased IL-4 and IL-5 production and increased IFN-α production. This study illustrates the viability of using chitosan as a delivery system for plasmid DNA in the treatment of asthma [74].

To address osteoarthritis and similar joint conditions, researchers have developed chitosan-hyaluronic acid hybrid nanoparticles loaded with plasmid DNA specifically targeting chondrocytes. When evaluating the transfection efficiency of chitosan nanoparticles using chondrocytes obtained from rabbit knee joint cartilage, it was observed that the presence of hyaluronic acid enhanced the transfection process. Furthermore, the effectiveness of transfection was influenced by factors such as plasmid concentration and pH levels (with the highest efficiency observed at pH 6.8). Importantly, the safety and utility of using chitosan for plasmid delivery in treating inflammatory joint ailments were demonstrated. This was achieved through achieving high transfection efficiency without causing any toxic effects on chondrocytes [75].

Another study generated a chitosan nanoparticle that was coated with polyethylenimine (PEI) and loaded with plasmid DNA that was intended to target synoviocytes. When compared to chitosan alone, it was discovered that the chitosan-PEI grafted system had a considerably greater transfection efficiency. In chondrocytes and synoviocytes isolated from rabbit knee cartilage, efficient and secure gene delivery is indicated by the maintenance of cell viability and the identification of plasmid DNA in the cell's nucleus after 4 hours. This study demonstrated the effectiveness of the chitosan system for delivering genes to treat inflammatory diseases like osteoarthritis [76].

Minocycline (MH-NPs), a tetracycline with antibacterial and anti-inflammatory actions, was designed for in situ delivery in the periodontal environment to enhance therapeutic effectiveness, according to Martin et al. With the right cytocompatibility profile, two different endocytic processes-clathrin-mediated endocytosis and macropinocytosis—can ingest MH-NPs. It has been found that it is possible to control autophagy while delivering drugs through the same endosomal/lysosomal pathway as periodontal infections, increasing the intracellular drug's efficacy. IL-1b, TNF, CXCL-8, and NFKB1 expression levels in Porphyromonas gingivalis LPS-stimulated cultures that were grown in the presence of MH-NPs were shown to be much lower. These nanoparticles hold promise for potential utilization in the treatment of periodontal disease, as they offer the capability to combine targeted intracellular drug delivery with potent anti-inflammatory characteristics [77].

A delivery strategy for acetyl-11-keto-β-boswellic acid (AKBA)-loaded o-carboxymethyl chitosan nanoparticles (AKBA-NP) was prepared by Ding et al. The tissue distribution investigation verified that AKBA-NPs had a higher efficacy of brain delivery than AKBA. The pharmacological experiments' findings demonstrated that AKBA-NPs have superior neuroprotection to AKBA in primary neurons under oxygen-glucose deprivation and in mice under middle cerebral artery occlusion models. Additionally, AKBA-NPs influence antioxidant and anti-inflammatory pathways more efficiently than AKBA by increasing heme oxygenase-1 expression and nuclear factor-kappa B expression while reducing 5-lipoxygenase and nuclear erythroid 2-related factor 2 expression [78].

Simvastatin (SVT) is administered orally on a platform of hybrid lecithin/chitosan nanoparticles (LCNs) as a drug delivery system for the management of neuroinflammatory disorders. Clementino et al. found that the interaction of nanoparticles with simulated nasal mucus reduced or retarded medication diffusion from nanoparticles. On the other hand, it was shown that the antimicrobial enzymes lysozyme and phospholipase A2, which are frequently found in nasal secretions, increased drug release from the nanocarrier. In fact, a 5-fold increase in drug release from LCNs was reported that was enzyme-triggered even in the presence of mucus. Additionally, nanoparticles coated with chitosan improved SVT penetration across a nasal epithelial model made of human cells (x11). The pro-inflammatory signaling was shown to be suppressed by SVT-LCNs more effectively than by the simple drug solution (75% for IL-6 and 27% for TNF-α at 10 M concentration for SVT-LCNs and SVT solution, respectively). Hybrid LCNs appear to be a perfect carrier for attaining nose-to-brain administration of poorly water-soluble medicines like SVT because of their biocompatibility and capacity to increase drug release and absorption at the biointerface [79].

Xiao et al. developed nanoparticles (NPs) by complexing CD98 siRNA (siCD98) with urocanic acid-modified chitosan (UAC) to facilitate the delivery of anti-inflammatory benefits. Subsequently, these nanoparticles were transfected into Raw 264.7 macrophages. No discernible cytotoxicity against colon-26 cells and Raw 264.7 macrophages as compared to control NPs made with Oligofectamine (OF) and siRNA. Cellular uptake tests showed that macrophages have a time-dependent trend of UAC/siRNA NP accumulation. UAC/siCD98 NPs with a weight ratio of 60:1 produced the most effective knockdowns of CD98 and the pro-inflammatory cytokine, TNF-α, according to additional in vitro gene silencing tests. In actuality, NPs' RNAi effectiveness was significantly higher than that of the NPs used as a positive control, OF/siCD98 [80].

Chitosan/small interfering RNA (siRNA) nanoparticles were used to inhibit tumor necrosis factor (TNF) expression in macrophages, contributing to the reduction of systemic and local inflammation in rheumatoid arthritis by Howard et al. In vitro, chitosan nanoparticles containing unmodified anti-TNF Dicer-substrate siRNA achieved a 66% TNF knockdown. Intraperitoneal (i.p.) administration of Cy3-labeled nanoparticles led to specific TNF knockdown (44%) in peritoneal macrophages in mice.

In mice with collagen-induced arthritis treated i.p. with anti-TNF-DsiRNA nanoparticles, joint swelling was halted, and histological examination showed minimal cartilage damage and inflammatory cell infiltration. Using 2'-O-Me-modified DsiRNA further reduced arthritic scores and decreased type I interferon (IFN) activation in macrophages in vivo compared to unmodified DsiRNA. This study demonstrates the potential of chitosan/siRNA nanoparticles as a therapeutic approach for managing rheumatoid arthritis by targeting TNF expression in macrophages [81].

Asif utilized the ionic gelation technique to create curcumin-loaded chitosan and sodium tripolyphosphate (STPP) nanoparticles to enhance curcumin's bioavailability. Four formulations were developed by varying STPP and chitosan concentrations, RPM, temperature, and pH of the chitosan solution. The NPs' characteristics, such as shape, drug-polymer compatibility, yield, particle size, encapsulation effectiveness, and release behavior, as well as their anti-inflammatory and anti-arthritic activities, were compared to curcumin and conventional drugs. The maximum yield achieved was 60%, with entrapment effectiveness ranging from 45% to 65%. The anti-inflammatory activity of curcumin NPs, measured by the HRB membrane stabilization method, was 59% higher than curcumin alone and 70% higher than the standard drug at a concentration of 600 µg/ml. The anti-arthritic activity, assessed using the protein denaturation method, was comparable to the standard drug and 66% higher than curcumin, reaching 70% improvement [82].

Alshehri and colleagues formulated a ternary nanoformulation (BPE-TOE-CSNPs NF) by encapsulating bee pollen extract (BPE) and thymol oil extract (TOE) within chitosan nanoparticles (CSNPs). The NF showed significant inhibition of inflammatory cytokine production (TNF-α and IL-6) in HepG2 and MCF-7 cells. Encapsulation of BPE and TOE in CSNPs enhanced treatment effectiveness, leading to cell cycle arrests in the S phase. The NF also induced strong apoptotic mechanisms by upregulating caspase-3 expression in cancer cells (2-fold in HepG2 and 9-fold in MCF-7). The novel NF showed increased expression of the P53 and caspase-9 apoptotic pathways. This nanoformulation holds promise as a potential treatment for cancer cells, particularly in MCF-7 cell lines [83].

Chitosan-alginate nanoparticles (NPs) exhibit direct antimicrobial activity against Propionibacterium acnes, the bacterium associated with acne development. Friedman et al. found that these NPs also demonstrated anti-inflammatory effects by inhibiting the production of inflammatory cytokines induced by P. acnes in human monocytes and keratinocytes. Furthermore, the chitosan-alginate NPs effectively encapsulated benzoyl peroxide (BP), an anti-acne drug, displaying stronger antimicrobial activity against P. acnes than BP alone. Importantly, the NPs showed reduced toxicity to eukaryotic cells. These findings suggest the potential of chitosan-alginate NPs as a promising approach for treating acne and inflammation while improving drug delivery and minimizing side effects [84].

In a study conducted by Bernela, nanoparticles were created using polycationic chitosan and polyanionic gum katira through an ionic complexation method. These nanoparticles effectively loaded glycyrrhizic acid with an encapsulation efficiency of 84.77% in an optimized formulation.

In vitro drug release experiments revealed sustained release of glycyrrhizic acid from the nanoparticles following zero-order kinetics, with 90.71% of the drug released over 12 hours. The release mechanism involved a combination of polymer matrix diffusion and erosion. Moreover, the study demonstrated that encapsulating glycyrrhizic acid in chitosan-katira gum nanoparticles significantly improved it's in vivo anti-inflammatory efficacy. This formulation overcame the limited bioavailability observed with other forms of glycyrrhizic acid, showcasing the potential of these nanoparticles as a promising approach for enhancing the therapeutic efficacy of glycyrrhizic acid in anti-inflammatory treatments [85].

In a study conducted by Tu et al., the potential anti-inflammatory effects of chitosan nanoparticles (CNP) were investigated on LPS-inflamed Caco-2 cells. The results of the study demonstrated that CNP mitigated the reduction in transepithelial electrical resistance (TEER) caused by LPS in Caco-2 cell monolayers. Furthermore, CNP exhibited significant inhibition of LPS-induced production of pro-inflammatory molecules such as TNF-α, MIF, IL-8, and MCP-1 in a dose-dependent manner. Additional analysis through PCR array assay unveiled that CNP downregulated the mRNA expression levels of TLR4 in LPS-inflamed Caco-2 cells. Moreover, CNP led to a reduction in cytoplasmic IκB-α degradation and nuclear NF-κB p65 levels in LPS-inflamed Caco-2 cells. These findings indicated that CNP played a role in suppressing the inflammatory response triggered by LPS by decreasing the permeability of the intestinal epithelial monolayer and restraining the secretion of pro-inflammatory cytokines in Caco-2 cells. This suppression was attributed, at least in part, to modulation of the NF-κB signaling pathway [86].

In a separate investigation, Reddy et al. undertook the creation of an innovative nanohybrid scaffold by incorporating curcumin (CUR) within chitosan nanoparticles (CSNPs) to enhance stability and solubility. The resulting CUR-CSNPs were subsequently integrated into a collagen scaffold, yielding a nanohybrid scaffold intended for improved tissue regeneration applications. The assessment of these nanoparticles revealed enhanced stability and solubility of CUR. The constructed nanohybrid scaffold exhibited favorable in vitro attributes, including improved water uptake, biocompatibility, and sustained drug release. In an in vivo analysis of wound closure, wounds treated with the nanohybrid scaffold displayed significantly accelerated contraction compared to wounds in the control and placebo scaffold groups ($p < 0.001$). This study proposes that the synergistic amalgamation of CUR (possessing anti-inflammatory and antioxidant properties), chitosan (serving as a sustained drug carrier and contributing wound healing properties), and collagen (established as a scaffold for wound healing) holds substantial potential in addressing various pathological aspects of diabetic wounds, thereby promoting more effective wound healing [87].

Nazanin Jabbari conducted a study involving the utilization of chitosan nanoparticles loaded with eugenol as a potent nano-herbal agent in the healing process of experimental neonatal rheumatoid arthritis (RA), with a comparison to methotrexate. The findings of the study demonstrated that both encapsulated eugenol by chitosan nanoparticles and methotrexate resulted in a notable reduction in the serum levels of malondialdehyde (MDA) and FOXO3 protein expression, as compared to the control group. Furthermore, the nanoparticle-based herbal agent and methotrexate exhibited a suppressive effect on the expression of TGF-β and MCP-1 genes, with a noteworthy positive correlation between MCP-1 and TGF-β. The study observed a decrease in inflammation, synovial hyperplasia, and pannus formation in rats with collagen-induced arthritis who were treated with the chitosan nanoparticle-loaded herbal agent and methotrexate. Thus, encapsulated eugenol in chitosan nanoparticles, along with methotrexate, has a protective effect against RA due to its immunomodulatory, anti-inflammatory, and antioxidant properties. Nano eugenol shows promise for treating autoimmune diseases like RA [88].

Abd El-Hameed looked into the impact of polydatin-loaded chitosan nanoparticles (POL-NPs) on early diabetic nephropathy in streptozotocin-induced diabetic rats. Blood glucose, blood glycosylated hemoglobin, serum insulin, renal function-related parameters, renal advanced glycation end products, and lipid peroxidation were all significantly higher in the diabetic group, while serum albumin levels and renal antioxidant enzyme activities were significantly lower. Nuclear factor-kappa B (NF-B) and cyclooxygenase-2 (Cox-2) mRNA expression was up-regulated in diabetic rat kidneys. Additionally, there was a drop in the anti-inflammatory cytokine (IL-10) and an increase in the pro-inflammatory cytokines (TNF-, IL-6, and IL-18) in the serum. POL and POL-NPs supplementation greatly lessened these effects and restored the healthy ratio of pro- to anti-inflammatory cytokines [89].

3.3 Exosomes for Anti-Inflammatory Effect

Exosomes are nanoscale vesicles derived from endosomes and are released by various cell types. They can be found in eukaryotic fluids like blood, urine, and cell culture medium, serving as vehicles for intercellular communication in both physiological and disease states. Exosomes typically have a diameter ranging from 30 to 100 nm and are formed through the reverse budding of multivesicular bodies, which fuse with the plasma membrane upon release [90-91].

Immunosuppressive dendritic cell (DC)-derived exosomes have been shown to possess potent and lasting immunosuppressive effects, similar to their parental DCs. These findings suggest that exosomes hold great potential as novel therapeutics for inflammatory conditions [92].

Exosomes, natural nanovesicles measuring 30-100 nm, are secreted by various cell types, including tumor cells, mesenchymal stem cells, and immune cells. They exhibit low cytotoxicity, non-immunogenicity, and endogenous properties, making them promising carriers for drug delivery [93].

Exosome-based drug delivery systems, leveraging the natural targeting property of exosomes derived from specific cell types to benefit disease treatment. However, its clinical application remains challenging due to the limitations associated with exosome isolation and purification, low loading efficiency of therapeutic cargoes, inadequate targeting capability, and rapid clearance from systemic circulation [94].

Exosomes have been explored by Zhuang et al., for the delivery of therapeutic agents in brain inflammatory-related diseases. For instance, exosomes encapsulating curcumin or a Stat3 inhibitor were noninvasively delivered to microglia cells via the intranasal route, resulting in protective effects against brain inflammation, experimental autoimmune encephalomyelitis, and brain tumor growth [95]. Turmeric-derived nanoparticles (TDNPs) have demonstrated excellent anti-inflammatory and antioxidant properties. In colitis models, orally administered TDNPs ameliorated colitis symptoms and regulated the expression of pro-inflammatory cytokines and antioxidant genes as reported by Liu. et al. [96].

A micelle-aided method has been developed by Lin et al. to efficiently load the anti-cancer agent imperialine into intact exosome-like vesicles (ELVs) for targeted drug delivery against inflammation and cancer. The ELVs were modified with an integrin $\alpha 3\beta 1$-binding octapeptide for tumor targeting, resulting in improved tumor accumulation and retention of imperialine with low systemic toxicity [97].

Monocyte-derived myeloid cells assume vital functions in inflammatory diseases. Curcumin, a natural polyphenol extracted from the rhizomes of Curcuma longa, exhibits anti-inflammatory, antineoplastic, and antioxidant properties. However, curcumin has low systemic bioavailability as described by Deng, et al. [98].

A study by Sun et al. published that, curcumin was incorporated with exosomes and injected as an exosome-curcumin complex into a mouse model. This complex protected lipopolysaccharide (LPS)-induced septic shock [99]. The complex formed by exosomes and curcumin augmented the anti-inflammatory impact of curcumin while evading immune responses and potential side effects. Specifically, in the context of colitis, the targeted delivery facilitated by exosomes allowed for precise and effective therapy [100].

Cai et al. discovered that exosomes derived from TGF-b1 gene-modified BMDC (TGF-b1-EXO) effectively inhibited the development of murine IBD induced by dextran sulfate sodium (DSS). The protective effect was dosage-dependent [101]. Wang et al. isolated exosomes from granulocytic myeloid-derived suppressor cells (G-MDSCs) and discovered that these G-MDSC exosomes effectively mitigated DSS-induced colitis and reinstated a balanced immune state within the intestines [102].

In a study by Wu et al. published molecularly engineered M2 macrophage-derived exosomes (m2 exo) were investigated. HAL@M2 Exo, which contained hexyl 5-aminolevulinate hydrochloride (HAL), exhibited excellent inflammation-tropism capability. In vitro and in vivo experiments demonstrated the anti-inflammatory effects of HAL@M2 Exo in atherosclerosis [103].

Yan et al. utilized exosomes as carriers and encapsulated dexamethasone sodium phosphate, modifying the surface with folic acid (FA)-polyethylene glycol (PEG)-cholesterol (Chol). In an in vitro study, the FPCExo/Dex drug delivery system exhibited an anti-inflammatory effect against RAW264.7 cells [104].

3.4　Liposomes

The self-assembling phospholipid bilayers that make up conventional liposomes, a type of nanomedicine, produce spherical vesicles with dimensions ranging from 30 nm to several micrometers. Hydrophilic pharmaceuticals can be placed into the aqueous interior of these vesicles, while hydrophobic medications can be put into the lipid bilayer. Conventional liposomes can have their properties and additional features customized by changing the way they are built. Lipidic nanoparticles were the first nanomedicine delivery system employed in clinical settings for various types of malignancies, and liposomes were the first closed bilayer phospholipid system [105].

The anti-inflammatory properties of liposomal nanoparticles are currently being researched [106]. In lipopolysaccharide-stimulated RAW 264.7 macrophages, Chiong et al. created a liposomal piroxicam formulation with cytoprotective and improved anti-inflammatory properties. Compared to cells treated with free-form piroxicam, those treated with liposome-encapsulated piroxicam showed better cell viability. Additionally, compared to piroxicam at an identical dose, the liposomal piroxicam formulation statistically inhibited pro-inflammatory mediators (such as nitric oxide, tumor necrosis factor-, interleukin-1, and prostaglandin E2) more effectively. Interleukin-

10, an anti-inflammatory cytokine, was also statistically significantly produced as a result of the liposome-encapsulated piroxicam.

In this study ApoE knockout mice were used by Duivenvoorden, to investigate the delivery of a statin using high-density lipoprotein nanoparticles. This delivery approach resulted in a decrease in the progression of atherosclerotic plaque inflammation [107].

Alternatively, liposomes have been employed for cholesterol vaccination by Lobatto et al. Rabbits were injected with liposomes containing 71% cholesterol and lipid A as an adjuvant, leading to the production of anti-cholesterol IgG and IgM antibodies. This approach resulted in reduced serum cholesterol levels, decreased antibody levels, decreased risk of atherosclerosis, and smaller plaque size [108].

Liposomes are also utilized for local drug delivery in the field of atherosclerosis by Van der Valk. In a rabbit model with atherosclerotic plaques induced by a high-cholesterol diet, PEGylated liposomes loaded with prednisolone phosphate were administered intravenously. This approach enabled the accumulation of prednisolone in the plaques, leading to effective treatment [109].

Interleukin-10 (IL-10), a cytokine known for its anti-inflammatory effects, has been administered using liposome cytokines.[110] Liposomes containing phosphatidylserine as a biomaterial carrier and encapsulating IL-10 showed significant anti-inflammatory and anti-obesity effects in an obese mouse model as described by Toita et al. PS, which mimics apoptotic cells [111], can induce a shift from inflammatory M1 macrophages to anti-inflammatory M2 macrophages and is specifically recognized by macrophages as reported by Nagata et al. [112].

In addition, liposomes modified with phosphatidylserine and DSPE-PEG2000-cRGDfK, termed apoptotic body liposomes, have been developed by Wu et al. These liposomes loaded with pioglitazone demonstrated the ability to target atherosclerotic plaque and exhibit anti-inflammatory effects by inhibiting M1 macrophage polarization and promoting M2 macrophage polarization [113,114].

Another application of liposomes involves modifying polyetheretherketone with dexamethasone and minocycline-loaded liposomes as published by Xu et al. In vitro and in vivo experiments demonstrated the enhanced anti-inflammatory, antibacterial, and Osseo integrative capacity of this hybrid nanoparticle, showing potential for clinical applications in orthopaedic and dental implants [115].

Liposomes containing PS have been used by Bagalkot, et al. to mimic cellular debris and have shown direct effects on anti-inflammatory cytokine production, highlighting their potential as anti-inflammatory and immunomodulatory agents [116].

3.5   Hydrogels/Nanogels

Polymer matrices are hydrogels that swell in water but do not dissolve. Nanogels are hydrogel nanoparticles that have undergone physical or chemical cross-linking with hydrophilic polymer strands. Because of these hydrogels' adaptable chemical and physical structure, outstanding mechanical properties, biocompatibility, and high water content, they have received a great deal of attention in the biomedical industry [117,118].

For the purpose of treating rabbits with osteochondral abnormalities, researchers created a collagen and resveratrol scaffold that is anti-inflammatory and cell-free [119].

Ratanavaraporn et al. investigated the use of a gelatin hydrogel with anti-inflammatory properties [120]. Gower et al. explored the application of a lactide-co-glycoside (PLG) scaffold as an anti-inflammatory carrier [121]. Nakamura et al. conducted a study on an anti-inflammatory hyaluronic acid hydrogel [122]. Kuskov et al. investigated the potential of self-assembled amphiphilic N-vinylpyrrolidone polymers (Amph-PVP) as a carrier for non-steroidal anti-inflammatory drugs of indomethacin [123]. These nanoparticles may be produced with diameters under 100 nm, narrow size distributions, high indomethacin contents (up to 35%), and high drug loading 24 efficiency (up to 95%). Indomethacin-loaded polymeric nanoparticles demonstrated greater 26-cell viability in cytotoxicity assays employing human embryonic stem cell-derived fibroblasts (EBF-H9) and hepatocellular carcinoma cells (HepG2) compared to free indomethacin at the same dose. According to the polymer's 28 molecular design and the Litchfield-Wilcoxon method, the median LD_{50} values for mice and rats were 55–70 mg/kg body weight. These carrier materials offer alternative options to nanogels and nanomedicines and have shown promise for their anti-inflammatory applications.

Wang et al. described a coating strategy that combined self-regulating anticoagulant and anti-inflammatory properties. Thrombin-responsive nanogels with anticoagulant and anti-inflammatory components were used in the technique. A thrombin-cleavable peptide was used to cross-link nanogels that contained the anticoagulant rivaroxaban. When exposed to thrombin from the environment, rivaroxaban was released, stopping the coagulation cascade. In order to provide antioxidant characteristics, the nanogels also contained the Tempol superoxide dismutase mimic. The polyphenol epigallocatechin gallate (EGCG) exhibited anti-inflammatory benefits via acting as a weak cross-linker and collaborating with Tempol. On biological valves and vascular stents, the coating was shown to efficiently resist coagulation and inflammation while encouraging reendothelialization, providing a promising method for creating functional coatings that resemble endothelium [124].

In a work reported by Aminu et al., PCL was used to create nanoparticles of the antibacterial medication triclosan. A hydrogel made of chitosan was directly loaded with the anti-inflammatory medication flurbiprofen. Nanogels (NG) were created when these nanoparticle and hydrogel systems were combined. In the chicken pouch model, the NG indicated bio-adhesiveness, while agar plate testing revealed antibacterial activity. The NG demonstrated dual antibacterial and anti-inflammatory properties, leading to remarkable therapeutic benefits in an in-vivo investigation on rats with experimental periodontitis [125].

The study conducted by Goel et al. investigated the anti-inflammatory activity of a nanogel formulation containing 3-acetyl-11-keto-β-boswellic acid (AKBA) against rat paw edema induced by carrageenan. The comparison between AKBA nanogel and AKBA gel of the same concentration showed significantly higher anti-inflammatory activity in the nanogel formulation [126].

In this study, Singka et al. prepared a nanogel composed of N-isopropyl acrylamide (NIPAM) and butyl acrylate (BA) was synthesized and loaded with MTX. The nanogel demonstrated de-swelling properties at temperatures ranging from 25 to 37 °C and exhibited enhanced MTX flux upon the addition of saturated aqueous Na_2CO_3. The study also observed reduced PGE2 levels when the MTX-loaded nanogel was applied, further highlighting its anti-inflammatory potential [127]. The MTX-loaded nanogel produced an MTX flow of 1.4± 0.3 ng cm-2 h-1, which increased to 3.1± 0.22 ng cm-2 h-1 upon the addition of saturated aqueous Na_2CO_3 (p 0.05) and showed de-swelling

Materials Research Forum LLC

https://doi.org/10.21741/9781644903858-6

by 7% over the range of 25-37° C. The relative lag times were 6 and 0 hours. Using aqueous Na_2CO_3 that was half-saturated, similar outcomes were attained. The silastic membrane was examined, and no permeability was found. When the MTX-loaded nanogel was administered, PGE2 levels decreased by 33% and by 57%, respectively, when saturated aqueous Na_2CO_3 has applied afterwards (p 0.01). PGE2 levels for water (control) and saturated aqueous Na_2CO_3 were identical (Figure 8).

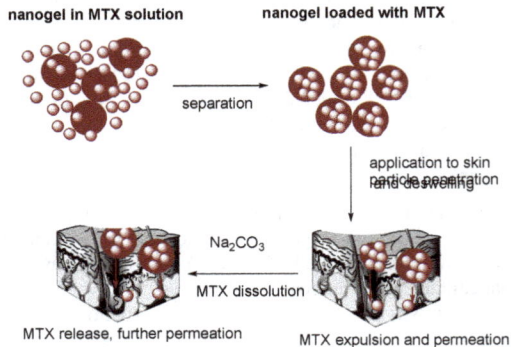

Figure 8. Use of MTX-loaded nanogel as drug delivery devise

Chen et al. developed a bactericidal polypeptide-crosslinked nanogel encapsulating tumor necrosis factor (TNF)-related apoptosis-inducing ligands (TRAIL). The TRAIL-nanogel treatment showed higher toxicity towards LPS-activated cells compared to naïve cells, and it effectively reduced bacterial numbers in circulation while prolonging survival in septic mice [128].

Li et al. described the fabrication of glycol chitosan/fucoidan nanogels loaded with the anti-inflammatory peptide KAFAK (GC/Fu@KAFAK NGs). The nanogels exhibited favorable characteristics such as an average size of 286.3 ± 5.0 nm and a positive surface charge. *In vitro* experiments on interleukin-1β (IL-1β)-stimulated rat chondrocytes demonstrated the nanogels' ability to inhibit inflammatory factors interleukin-6 (IL-6) and tumor necrosis factor-alpha (TNF-α) and enhance the expression of chondrogenic markers type II collagen, aggrecan, and Sox9. Furthermore, in a rat osteoarthritis model, the injection of the nanogels GC/Fu@ KAFAK NGs reduced glycosaminoglycan loss and suppressed inflammatory cytokine release [129]. Both in vitro and in vivo tests revealed good biocompatibility (Figure 9).

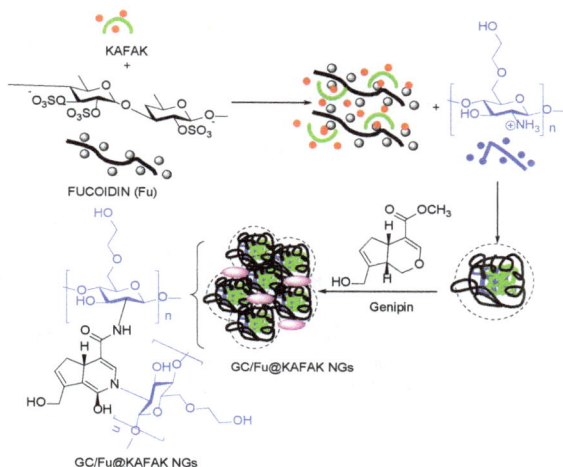

Figure 9. KAFAK-loaded glycol chitosan/fucoidan nanogels

Valentino et al. focused on the preparation of micro/nanogels using alginate (ALG) and spermidine (SP) as a cross-linking agent and an antioxidant/anti-inflammatory compound, respectively. The addition of trehalose as a cryoprotectant agent was considered to maintain stability during freeze-drying. In vitro studies on Schwann cells demonstrated the antioxidant and anti-inflammatory properties of SP, highlighting its role in the formation of nanogels [130].

Nucleic acid nanogels were combined with a liposomal vesicle to develop a hybrid system (Van-DNL) for antibiotic delivery in a work by Obuobi, et al. Antibiotics are immobilized by non-covalent electrostatic interactions with polyanionic DNA and cationic cargos, allowing for precise temporal release against intracellular S. aureus. Van-DNL was identified through in vitro testing as a stable homogeneous formulation with improved vancomycin loading efficiency and circular morphology. In contrast to controls, this hybrid system showed persistent vancomycin release. When utilized to treat S. aureus-infected macrophages, the hybrid system likewise showed a dose-dependent reduction in intracellular bacterial load over 24 h and a potent anti-inflammatory effect. It was discovered that 48 hours of exposure to Van-DNL had very little effect on cellular toxicity [131].

Yeo et al. reported the development of a tannic acid-based nanogel with high nitric oxide-scavenging properties for the treatment of rheumatoid arthritis. The nanogel, known as polymeric phenylboronic acid-tannic acid nanogel (PTNG), exhibited significant antioxidant effects and strong anti-inflammatory effects in an induced inflammation model. In a zymosan-induced peritonitis mouse model, PTNG successfully reduced neutrophil recruitment and pro-inflammatory cytokines, indicating effective alleviation of inflammation [132].

Shah et al. conducted a study in 2012 where they developed a nanogel system for delivering spantide II (SP) and ketoprofen (KP), which are anti-inflammatory drugs, through the skin. The system consisted of poly-(lactide-co-glycolic acid) and chitosan-based bilayered nanoparticles

Nanomaterials in Biological Systems   Materials Research Forum LLC
Materials Research Foundations 185 (2026) 112-153   https://doi.org/10.21741/9781644903858-6

(NPs) modified with oleic acid (NPSO) to get skin permeating nanogel system (SPN). Hydroxypropyl methylcellulose (HPMC) and Carbopol were used to prepare the nanogels. To evaluate the efficacy of SPN, allergic contact dermatitis (ACD) and a psoriatic plaque-like model were utilized [133,134]. Studies on in vitro permeation showed that SP-SPN or SPKP-SPN increased SP deposition in the epidermis and dermis by 8.5 and 9.5 times more than SP-gel. Furthermore, compared to KP-gel, the deposition of KP for KPSPN or SP-KP-SPN was higher in the epidermis and dermis by 9.75 and 11.55 folds, respectively. Similar to KPgel, KP-SPN or SP-KP-SPN had KP permeation that was 9.92 times more than KPgel. The PASI score and TEWL values in the psoriatic plaque-like model; ear thickness in the ACD model; and the expression of IL-17 and IL-23; were considerably lower (p 0.001) for SPN compared to the control gel.

Gutowski et al. transformed the naturally occurring antagonist of the IL-1R, IL-1RA, into a PEG-maleimide hydrogel with on-demand protease-sensitive release. The use of this technology as a covering reduced inflammation caused by the implantation of brain electrodes. RT-PCR analysis revealed only small variances, with elevations in IL-6, MMP-13, and ciliary neurotrophic factor (CNTF) being the only modifications observed, in contrast to immunofluorescent labelling of inflammatory cell infiltration, which revealed no differences between samples [135,136].

In 2018, Giulbudagian et al. focused on achieving sustained delivery of a thermoresponsive nanogel (tNG) and encapsulating the anti-TNFα fusion protein etanercept (ETR) without altering its structure. The researchers successfully delivered ETR to human skin by applying ETR-loaded tNGs to inflammatory skin equivalents or tape-striped human skin. The release of ETR from the tNGs was triggered by temperature, and it exhibited anti-inflammatory effects [137].

Fariba Esmaeili investigated the anti-inflammatory and anti-nociceptive effects of nanoemulsion-based gels containing essential oils in an animal model. The study prepared nanoemulsions of cinnamon and clove essential oils with droplet sizes of 28 ± 6 nm and 12 ± 3 nm, respectively. Carboxymethyl cellulose was added to gelify the nanoemulsions. The findings of the paw edema and formalin tests demonstrated that the nanogel formulations exhibited significant anti-inflammatory and anti-nociceptive effects [138].

Lin et al. conducted a study in 2022 where they developed long-lasting eye drops using lysine-carbonized nanogels (Lys-CNGs) derived from carbonized nanogels (CNGs) via lysine hydrochloride (Lys) pyrolysis. The Lys-CNGs exhibited high biocompatibility with corneal epithelial cells both in vitro and in vivo, indicating their safety as eye drops. The study also demonstrated that a single dose of Lys-CNGs effectively alleviated the signs of dry eye syndrome (DES) in a rabbit model within four days, showing better therapeutic effects compared to multiple treatments of a higher concentration of cyclosporine A [139]. They manufactured nanocarriers that imitate the arthritic extracellular environment. Argan oil-based emulgel is used to create biocompatible, biodegradable ketoprofen-loaded chitosan-chondroitin sulfate (CHS-CS) nanocarriers for transdermal applications with natural substances. This enhances the formulation's anti-inflammatory impact and speeds up skin penetration. These nanocarriers contained ketoprofen, which released 77% of its medication over the course of 72 hours at pH 7.4 and showed an encapsulation effectiveness of over 76%. The simple diffusion method used by CHS-CS nanoparticles to deliver drugs. The nanoparticle-loaded argan oil emulgel significantly boosted the epidermal permeation of ketoprofen when compared to commercial gel (p 0.05).

Whitmire et al. reported in 2012 that self-assembling copolymer nanoparticles (NPs) can prolong the retention time of IL-1 receptor antagonist (IL-1Ra) when it is covalently conjugated on the surface of the NPs and delivered locally. IL-1Ra is a natural protein inhibitor of IL-1 [140].

With a protein tethering moiety for surface covalent attachment of IL-1Ra protein, this block copolymer self-assembles into 300 nm-diameter particles, maintaining protein bioactivity in vitro. Using in vivo imaging over 14 days, the capacity of nanoparticles to maintain IL-1Ra in the rat stifle joint was assessed. Without causing degenerative changes in cartilage structure or composition, IL-1Ra-tethered nanoparticles considerably extended the retention period of IL-1Ra in the rat stifle joint for 14 days with an improved IL-1Ra half-life (3.01 days) compared to that of soluble IL-1Ra (0.96 days) (Figure 10).

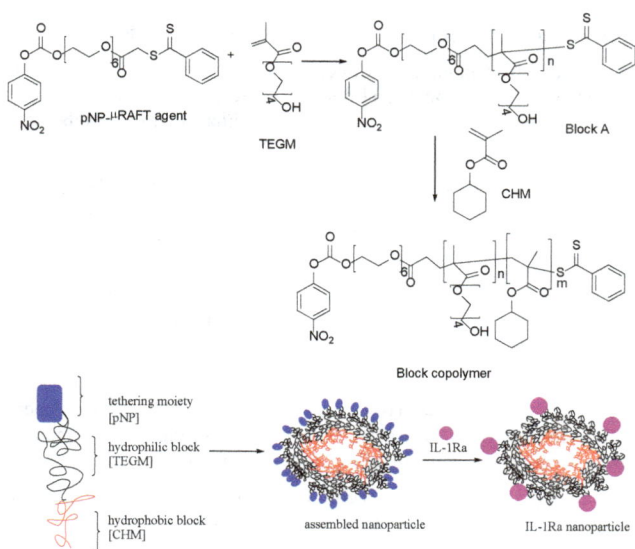

Figure 10. IL-1Ra-tethered nanoparticles for anti-inflammatory effect

3.6 Miscellaneous

Webber et al. used dexamethasone on a peptide nanofiber gel to modulate the inflammatory response. In vitro experiments revealed that dexamethasone delivered via the peptide nanofiber gel inhibited NF-kB activity in LPS-treated THP-1 monocytes. An in vivo subcutaneous study discovered that dexamethasone-loaded gels reduced the presence of ROS and inflammatory cells after 3 and 21 days when compared to a non-dexamethasone-loaded peptide nanofiber gel [141].

The administration of a therapeutic siRNA via a lipid nanoparticle by Leuschner et al. reduced the encroachment of pro-inflammatory monocytes into inflamed tissue. A siRNA targeting the chemokine receptor CCR2 was administered systemically, and it was shown to reduce the size of the infarct in a model of MI, the number of inflammatory cells in atherosclerotic lesions, the survival of pancreatic islet allografts, and tumour volume. In this scenario, pro-inflammatory macrophages are used to demonstrate siRNA's therapeutic potential as a tool for modifying cell phenotype [142].

According to research by Park et al., urethane acrylate non-ionomer (UAN) nanoparticles were functionalized with a targeting moiety to improve localisation to tumors. Nanoparticles with an ICAM-1 targeting ligand attached improved localisation to tumors and inflamed tissue. Compared to those without a targeting ligand, UAN nanoparticles loaded with paclitaxel reduced the amount of the tumor. This demonstrates the viability of using nanoparticles with a targeting moiety to target inflammation. This technology can be modified to carry several medications and can also be used in conjunction with different ligands to target different regions [143].

4. Mechanism of Anti-Inflammatory Effect with Nanomaterials

The mechanism of the anti-inflammatory effect with nanomaterials can vary depending on the specific characteristics of the nanomaterial and the targeted inflammatory response [144,145]. However, here are some general mechanisms by which nanomaterials can exhibit anti-inflammatory effects:

a. *Modulation of immune response*: Nanomaterials can interact with immune cells and modulate their activity. They may promote the production of anti-inflammatory cytokines while suppressing the release of pro-inflammatory cytokines. This modulation can help regulate the immune response and reduce inflammation.

b. *Targeted drug delivery*: Nanomaterials can encapsulate or conjugate anti-inflammatory drugs and deliver them to the site of inflammation. This targeted drug delivery approach allows for localized and sustained release of the therapeutic agents, maximizing their efficacy while minimizing systemic side effects.

c. *Scavenging reactive oxygen species* (ROS): Some nanomaterials possess inherent antioxidant properties and can scavenge ROS, which are known to contribute to inflammation. By reducing ROS levels, these nanomaterials can help alleviate inflammation and oxidative stress.

d. *Inhibition of inflammatory signaling pathways*: Certain nanomaterials can interfere with specific signaling pathways involved in inflammation, such as NF-κB (nuclear factor kappa-light-chain-enhancer of activated B cells), which regulates the expression of many pro-inflammatory genes. By inhibiting these pathways, nanomaterials can downregulate the production of inflammatory mediators.

e. *Immune tolerance induction*: Nanomaterials can be engineered to induce immune tolerance, which involves training the immune system to become less reactive to specific antigens. By promoting immune tolerance, nanomaterials can reduce exaggerated immune responses and dampen chronic inflammation.

It is important to note that the exact mechanism of the anti-inflammatory effect can vary depending on factors such as the specific nanomaterial composition, size, surface properties, and the target tissue or disease. The field of nanomedicine is continually advancing, and researchers are exploring various nanomaterials and their unique properties for the development of effective anti-inflammatory therapies (Figure 11).

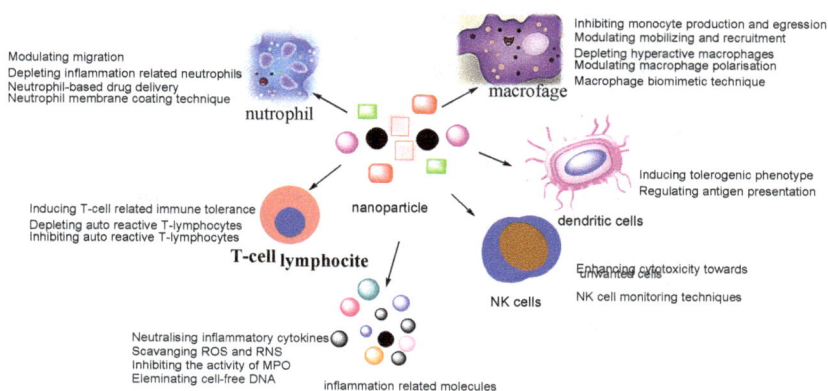

Figure 11. *Plausible mechanisms of anti-inflammatory effect with nanoparticles*

Conclusions

The progress in nanotechnology has played a significant role in exploring potential anti-inflammatory activity and paving the way for future applications of nanomedicine in treating various inflammatory diseases such as inflammatory bowel disease, rheumatoid arthritis, osteoarthritis, wound healing, and sepsis. Several types of organic nanoparticles have shown promise in combatting inflammation. Among them, biodegradable and biocompatible organic nanoparticles like PLGA and hydrogel have been thoroughly investigated. Mechanism studies have also been discussed in detail in related research. Organic nanoparticles offer a targeted drug delivery approach, enabling the loading of anti-inflammatory drugs to be released specifically at the site of inflammation. This strategy holds the potential to reduce side effects and enhance the therapeutic efficacy of medications, particularly in diseases like inflammatory bowel disease (IBD). For instance, nanoparticles possess the ability to scavenge reactive oxygen species (ROS) or deliver growth factors, which can aid in tissue repair and decrease inflammation for conditions like rheumatoid arthritis (RA) and osteoarthritis (OA). Additionally, organic nanoparticles can serve as carriers for growth factors, antimicrobial agents, and other molecules that facilitate tissue regeneration and reduce inflammation at the site of wounds. This property proves particularly valuable for promoting wound healing. In cases of sepsis, a severe and potentially life-threatening condition caused by an overwhelming response to infection, organic nanoparticles can function as carriers for antibiotics, leading to improved drug delivery to the infection site. Overall, the utilization of organic nanoparticles in drug delivery and therapeutic interventions holds great promise for managing inflammatory diseases and improving patient outcomes.

References

[1] M.Y. Putra, T. Murniasih, Marine soft corals as source of lead compounds for anti-inflammatories. J Coast Life Med. 4 (2016), 73–77. 10.12980/jclm.4.2016j5-226

[2] L. Ferrero-Miliani, O. H. Nielsen, P. S. Andersen, and S. E. Girardin. Chronic inflammation: the importance of NOD2 and NALP3 in interleukin-1 beta generation. Clin. Exp. Immunol. 147 (2007), 227–235. DOI: 10.1111/j.1365-2249.2006.03261.x

[3] J. Burisch, P. Lakatos, M. Martinato, J. Tine, The burden of inflammatory bowel disease in Europe. J Crohns Colitis. 7(2013), 322-337. DOI: 10.1016/j.crohns.2013.01.010

[4] Y. C. Yang, M. K. McClintock, M. Kozloski, and T. Li. Social isolation and adult mortality: The role of chronic inflammation and sex differences. J. Health Soc. Behav. 54 (2013), 183-203. DOI: 10.1177/0022146513485244

[5] R. S. Cotran, V. Kumar, T. Collins, S. L. Robbins. Robbins pathologic basis of disease. Philadelphia: Saunders (1999). ISBN 0-7216-0187-1.

[6] B.Y. Hanaoka, M.P. Ithurburn, C.A. Rigsbee, et al. Chronic inflammation in rheumatoid arthritis and mediators of skeletal muscle pathology and physical impairment: a review. Arthritis Care Res. 71(2019), 173–177. DOI: 10.1002/acr.23775

[7] G. R. Geovanini, P. Libby, Atherosclerosis and inflammation: Overview and updates. Clin. Sci.132 (2018), 1243–1252. 10.1042/CS20180303

[8] G. G. Biondi-Zoccai, A. Abbate, G. Liuzzo, L. M. Biasucci, Atherothrombosis, inflammation, and diabetes. J. Am. Coll. Cardiol. 41(2003), 1071–1077. DOI: 10.1016/S0735-1097(03)00088-3

[9] H. Akiyama, S. Barger, S. Barnum, et al. Inflammation and Alzheimer's disease. Neurobiol. Aging. 21(2000), 383–421. DOI: 10.1016/S0197-4580(00)00124-X

[10] R. K. Cross, K. T. Wilson. Nitric oxide in inflammatory bowel disease. Inflamm. Bowel Dis. 9 (2003), 179–189. DOI: 10.1097/00054725-200305000-00006

[11] L. M. Coussens, Z. Werb, Inflammation and cancer. Nature, 420 (2002), 860. DOI: 10.1038/nature01322

[12] M.C. Siracusa, B.S. Kim, J.M. Spergel, D. Artis. Basophils and allergic inflammation. J. Allergy Clin. Immunol. 132(2013), 789–801. DOI: 10.1016/j.jaci.2013.07.046

[13] P.H. Leliefeld, C.M. Wessels, L.P. Leenen, L. Koenderman, J. Pillay. The role of neutrophils in immune dysfunction during severe inflammation. Critical Care; 20 (2016):73. DOI: 10.1186/s13054-016-1250-4

[14] Geering B., Stoeckle C., Conus S., Simon H.-U. Living and dying for inflammation: neutrophils, eosinophils, basophils. Trends Immunol. 34(2013), 398–409. DOI: 10.1016/j.it.2013.03.004

[15] M. Rai, S. Deshmukh, A. Ingle, A. Gade, Silver nanoparticles: the powerful nano weapon against multidrug-resistant bacteria. J. Appl. Microbiol. 112(2012), 841–852. DOI: 10.1111/j.1365-2672.2012.05253.x

[16] W. H. Gerwick, B. S. Moore, Lessons from the past and charting the future of marine natural products drug discovery and chemical biology. ChemBiol. 19(2012), 85–98. DOI: 10.1016/j.chembiol.2011.12.014

[17] N Shady, M. Fouad, K. M. Salah, Schirmeister T, Abdelmohsen U. Natural Product Repertoire of the Genus Amphimedon. Mar. Drugs, 17(2019), 19. DOI: 10.3390/md17010019

[18] F.S. Alves, J.N. Cruz, I.N. de Farias Ramos, D.L. do Nascimento Brandão, R.N. Queiroz, G.V. da Silva, G.V. da Silva, M.F. Dolabela, M.L. da Costa, A.S. Khayat, J. de Arimatéia Rodrigues do Rego, D. do Socorro Barros Brasil, Evaluation of Antimicrobial Activity and Cytotoxicity Effects of Extracts of Piper nigrum L. and Piperine, Separations. 10 (2023). DOI: 10.3390/separations10010021

[19] I.G. Rodrigues, M.G. Miguel, W. Mnif, A brief review on new naturally occurring cembranoid diterpene derivatives from the soft corals of the genera Sarcophyton, Sinularia, and Lobophytum since 2016. Molecules, 24(2019), 781. DOI: 10.3390/molecules24040781

[20] Q. Zhang, L.-F. Liang, Z.-H. Miao, B. Wu, Y.-W. Guo, Cytotoxic polyhydroxylated steroids from the South China Sea soft coral Lobophytum sp. Steroids, 141(2019), 76–80. DOI: 10.1016/j.steroids.2018.11.015

[21] K.-K. Gong, X.-L. Tang, G. Zhang, et al. Polyhydroxylated steroids from the South China Sea soft coral Sarcophyton sp. and their cytotoxic and antiviral activities. Mar. Drugs, 11(2013),4788–4798. DOI: 10.3390/md11124788

[22] I. Guibert, I. Bonnard, X. Pochon, et al. Differential effects of coral giant clam assemblages on biofouling formation. Sci Rep. 9 (2019), 2675. DOI: 10.1038/s41598-019-39268-1

[23] M Oray, A. K. Samra, N. Ebrahimiadib, H. Meese, C.S. Foster, Long term side effects of glucocorticoids. Expert Opin Drug Saf. 15 (2016), 457–465. DOI: 10.1517/14740338.2016.1140743

[24] C. Sostres, C.J. Gargallo, M.T. Arroyo, A. Lanas, Adverse effects of non-steroidal anti-inflammatory drugs (NSAIDs, aspirin and coxibs) on upper gastrointestinal tract. Best Pract. Res. Clin. Gastroenterol. 24 (2010), 121–132. DOI: 10.1016/j.bpg.2010.03.002

[25] M.H. Sarfraz, M. Zubair, B. Aslam, A. Ashraf, M.H. Siddique, S. Hayat, J.N. Cruz, S. Muzammil, M. Khurshid, M.F. Sarfraz, A. Hashem, T.M. Dawoud, G.D. Avila-Quezada, E.F. Abd_Allah, Comparative analysis of phyto-fabricated chitosan, copper oxide, and chitosan-based CuO nanoparticles: antibacterial potential against Acinetobacter baumannii isolates and anticancer activity against HepG2 cell lines, Front. Microbiol. 14 (2023). DOI:10.3389/fmicb.2023.1188743.

[26] J.N. Cruz, S. Muzammil, A. Ashraf, M.U. Ijaz, M.H. Siddique, R. Abbas, M. Sadia, Saba, S. Hayat, R.R. Lima, A review on mycogenic metallic nanoparticles and their potential role as antioxidant, antibiofilm and quorum quenching agents, Heliyon. 10 (2024). DOI:10.1016/j.heliyon.2024.e29500

[27] W. K. Leutwyler, S. L. Bürgi, H. Burgl, Semiconductor clusters, nanocrystals, and quantum dots, Science 271(1996), 933-7. DOI:10.1126/science.271.5251.933

[28] S. Gelperina, K. Kisich, M.D. Iseman, L. Heifets, The potential advantages of nanoparticle drug delivery systems in chemotherapy of tuberculosis, Am J Respir Crit Care 172(2005),1487-1490. DOI: 10.1164/rccm.200504-613PP

[29] Y. Cui, Q. Wei, H. Park, C.M. Lieber, Nanowire Nanosensors for Highly Sensitive and Selective Detection of Biological and Chemical Species, Science 293 (2001), 1289-92. DOI: 10.1126/science.1062711

[30] B. Fubini, Surface reactivity in the pathogenic response to particulates. Environ Health Perspect. 105(1997),1013-20. DOI: 10.1289/ehp.97105s51013

[31] X. Li, L. Wang, Y. Fan, Q. Feng, F. Cui, Biocompatibility and toxicity of nanoparticles and nanotubes, J. Nanomat. 2012(2012), 548389. DOI: 10.1155/2012/548389

[32] S. Bera, D. Mondal, Antibacterial Efficacies of Nanostructured Aminoglycosides. ACS Omega 7 (2022), 4724–4734. https://doi.org/10.1021/acsomega.1c04399

[33] G.A. Silva, Nanotechnology approaches to crossing the blood-brain barrier and drug delivery to the CNS. BMC Neurosci. 9(2008), S4. doi: 10.1186/1471-2202-9-S3-S4

[34] P. Lutwyche, C. Cordeiro, D.J. Wiseman, M. St-Louis, M. Uh, M.J. Hope, M.S. Webb, B.B. Finlay, Intracellular delivery and antibacterial activity of gentamicin encapsulated in pH-sensitive liposomes, Antimicrob. Agents Chemother. 42(1998), 2511. doi: 10.1128/AAC.42.10.2511

[35] S. Mitragotri, S. Patrick, Organic nanoparticles for drug delivery and imaging. MRS Bulletin 39(2014), 219–223. doi: 10.1557/mrs.2014.11

[36] O. Farokhzad, J. Cheng, B. Teply, I. Sherifi, S. Jon, P. Kantoff, J. Richie, R. Langer, Targeted nanoparticle-aptamer bioconjugates for cancer chemotherapy in vivo, Proc. Natl. Acad. Sci. U.S.A. 103 (2006), 6315. doi: 10.1073/pnas.0601755103

[37] J. B. Lin, S. Poh, and A. Panitch, Controlled release of anti-inflammatory peptides from reducible thermosensitive nanoparticles suppresses cartilage inflammation. Nanomedicine 12(2016), 2095–2100. doi: 10.1016/j.nano.2016.05.010

[38] M. C. Chang, Y. J. Kuo, K. H. Hung, C.-L. Peng, K.-Y. Chen, and L.-K. Yeh, Liposomal dexamethasone-moxifloxacin nanoparticle combinations with collagen/gelatin/alginate hydrogel for corneal infection treatment and wound healing. Biomed. Mater. 15 (2020), 055022. doi: 10.1088/1748-605X/ab9510

[39] S Browne, and A. Pandit, Biomaterial-mediated modification of the local inflammatory environment. Front. Bioeng. Biotechnol. 3(2015), 67. doi: 10.3389/fbioe.2015.00067

[40] Y. Z. Zhang, and Y. Y. Li, Inflammatory bowel disease: pathogenesis. World J. Gastroenterol. 20(2014), 91–99. doi: 10.3748/wjg.v20.i1.91

[41] A. Lamprecht, N. Ubrich, H. Yamamoto, U. Schäfer, H. Takeuchi, P. Maincent, et al. Biodegradable nanoparticles for targeted drug delivery in the treatment of inflammatory bowel disease. J. Pharmacol. Exp. Ther. 299 (2001), 775–781. PMID: 11602694.

[42] Y. Meissner, Y. Pellequer, A. Lamprecht, Nanoparticles in inflammatory bowel disease: particle targeting versus pH-sensitive delivery. Int. J. Pharm. 316(2006), 138-43. DOI: 10.1016/j.ijpharm.2006.01.032

[43] A. Lamprecht, U. Scha"fer, C.M. Lehr, Size-dependent bioadhesion of micro- and nanoparticulate carriers to the inflamed colonic mucosa. Pharm Res 18(2001),788–93. DOI: 10.1023/A:1011032328064

[44] Y. Shi, F. Xie, P. Rao, H. Qian, R. Chen, H. Chen, et al. TRAIL expressing cell membrane nanovesicles as an anti-inflammatory platform for rheumatoid arthritis therapy. J. Control Release 320(2020b), 304–313. DOI 10.1016/j.jconrel.2020.01.054

[45] Y. Zhou, W. Xu, and A. Shao, Application prospect of mesenchymal stem cells in the treatment of sepsis. Crit. Care Med. 48(2020), e634. DOI 10.1097/CCM.0000000000004341

[46] Y. Wang, J.H. Zhang, J. Sheng and A. Shao, Immunoreactive Cells After Cerebral Ischemia. Front. Immunol. 10(2019), 2781. DOI 10.3389/fimmu.2019.02781

[47] I. Bala, S. Hariharan, and M. N. Kumar, PLGA nanoparticles in drug delivery: the state of the art. Crit. Rev. Ther. Drug Carrier Syst. 21(2004), 387–422. DOI 10.1615/critrevtherdrugcarriersyst.v21.i5.20

[48] M. Mir, N. Ahmed, and A. U. Rehman, Recent applications of PLGA based nanostructures in drug delivery. Colloids Surf. B Biointerfaces 159(2017), 217–231. DOI 10.1016/j.colsurfb.2017.07.038

[49] P. T. Thevenot, A. M. Nair, J. Shen, P. Lotfi, C.-Y. Ko, and L. Tang, The effect of incorporation of SDF-1alpha into PLGA scaffolds on stem cell recruitment and the inflammatory response. Biomaterials 31(2010), 3997–4008. doi: 10.1016/j.biomaterials.2010.01.144.

[50] G. Fredman, N. Kamaly, S. Spolitu, J. Milton, D. Ghorpade, R. Chiasson, et al. Targeted nanoparticles containing the proresolving peptideAc2-26 protect against advanced atherosclerosis in hypercholesterolemic mice. Sci. Transl. Med. 7(2015), 275ra20. DOI 10.1126/scitranslmed.aaa1065

[51] W. Gu, C. Wu, J. Chen, and Y. Xiao, Nanotechnology in the targeted drug delivery for bone diseases and bone regeneration. Int. J. Nanomed. 8(2013), 2305–2317. DOI: 10.2147/IJN.S44393

[52] K. Feng, H. Sun, M. A. Bradley, E. J. Dupler, W. V. Giannobile, and P. X. Ma, Novel antibacterial nanofibrous PLLA scaffolds. J. Control Release 146(2010), 363–369. DOI: 10.1016/j.jconrel.2010.05.035

[53] K. T. Peng, C. F. Chen, I. M. Chu, Y.-M. Li, W.-H. Hsu, R. W.-W. Hsu, et al. Treatment of osteomyelitis with teicoplanin-encapsulated biodegradable thermosensitive hydrogel nanoparticles. Biomaterials 31(2010), 5227–5236. DOI: 10.1016/j.biomaterials.2010.03.027

[54] R. M. Trujillo-Nolasco, E. Morales-Avila, B. E. Ocampo-Garcia, G. Ferro-Flores, B. V. Gibbens-Bandala, A. Escudero-Castellanos, et al. Preparation and in vitro evaluation of radio-labeled HA-PLGA nanoparticles as novel MTX delivery system for local treatment of rheumatoid arthritis. Mater. Sci. Eng. C Mater. Biol. Appl. 103(2019), 109766. DOI: 10.1016/j.msec.2019.109766

[55] T. Yazeji, B. Moulari, A. Beduneau, V. Stein, D. Dietrich, Y. Pellequer, et al. Nanoparticle-based delivery enhances anti-inflammatory effect of low molecular weight heparin in experimental ulcerative colitis. Drug Deliv. 24(2017), 811–817. DOI: 10.1080/10717544.2017.1324530

[56] K. Tahara, S. Samura, K. Tsuji, H. Yamamoto, Y. Tsukada, Y. Bando et al. Oral nuclear factor-kappaB decoy oligonucleotides delivery system with chitosan modified poly(D,L-lactide-co-glycolide) nanospheres for inflammatory bowel disease. Biomaterials 32(2011), 870–878. DOI: 10.1016/j.biomaterials.2010.09.034

[57] X. Zhang, C. Yu, Xushi, C. Zhang, T. Tang, K. Dai, Direct chitosan-mediated gene delivery to the rabbit knee joints in vitro and in vivo. Biochem. Biophys. Res. Commun. 341(2006), 202–208. DOI: 10.1016/j.bbrc.2005.12.171

[58] D.-W. Lee, S. A. Shirley, R. F. Lockey, S. S. Mohapatra, Thiolated chitosan nanoparticles enhance anti-inflammatory effects of intranasally delivered theophylline. Respiratory Research 7, Article number: 112 (2006). DOI: 10.1186/1465-9921-7-112

[59] J. Zhang, C. Tang, and C. Yin, Galactosylated trimethyl chitosan-cysteine nanoparticles loaded with Map4k4 siRNA for targeting activated macrophages. Biomaterials 34 (2013), 3667–3677. DOI: 10.1016/j.biomaterials.2013.01.079

[60] J. M. Soni, M. N. Sardoiwala, S. Roy Choudhury, S. S. Sharma, S. Karmakar. Melatonin-loaded chitosan nanoparticles endows nitric oxide synthase 2 mediated anti-inflammatory activity in inflammatory bowel disease model. Materials Science and Engineering: C, 124 (2021), 112038. 10.1016/j.msec.2021.112038

[61] D. K. Podolsky, Inflammatory bowel disease. New Engl J Med, 347 (2002), 417–29. doi:10.1056/NEJMra020831

[62] D. Pertuit, B. Moulari, T. Betz, et al. 5-amino salicylic acid bound nanoparticles for the therapy of inflammatory bowel disease. J. Control Rel. 123 (2007), 211–18. doi:10.1016/j.jconrel.2007.08.008

[63] P. Tang, Q. Sun, L. Zhao, H. Pu, H. Yang, S. Zhang, R. Gan, N. Gan, H. Li. Mesalazine/hydroxypropyl-β-cyclodextrin/chitosan nanoparticles with sustained release and enhanced anti-inflammation activity. Carbohydrate Polymers, 198 (2018), 418-425. doi: 10.1016/j.carbpol.2018.06.106

[64] A. I. Barbosa, S. A. C. Lima, S. Reis. Development of methotrexate loaded fucoidan/chitosan nanoparticles with anti-inflammatory potential and enhanced skin permeation. Int. J. Biolog. Macromol. 124(2019), 1115-1122. doi: 10.1016/j.ijbiomac.2018.12.014

[65] A. A. Öztürk, H. T. Kıyan. Treatment of oxidative stress-induced pain and inflammation with dexketoprofen trometamol loaded different molecular weight chitosan nanoparticles: Formulation, characterization and anti-inflammatory activity by using in vivo HET-CAM assay. Microvasc. Res. 128 (2020), 103961. doi:10.1016/j.mvr.2019.103961

[66] Y. Haitao, C. Yifan, S. Mingchao and H. Shuaijuan, A Novel Polymeric Nanohybrid Antimicrobial Engineered by Antimicrobial Peptide MccJ25 and Chitosan Nanoparticles

Materials Research Forum LLC
https://doi.org/10.21741/9781644903858-6

Exerts Strong Antibacterial and Anti-Inflammatory Activities. Front. Immunol. 12, (2022) 811381. doi:10.3389/fimmu.2021.811381

[67] A. S. Hirva, P. P. Rakesh. Statistical modeling of zaltoprofen loaded biopolymeric nanoparticles: Characterization and anti-inflammatory activity of nanoparticles loaded gel. Int. J. Pharma. Investig. 5(2015), 120-27. doi:10.4103/2230-973X.147229

[68] K. Haliza, H. Zahid, C. L. Tay. Chitosan Nanoparticles as a percutaneous drug delivery system for hydrocortisone. J. Nanomater. 1 (2012), 1-11. doi:10.1155/2012/372725

[69] D. A. Mehmet, U. Nagihan, Y. Fatma, et al. Ketorolac tromethamine loaded chitosan nanoparticles as a nanotherapeutic system for ocular diseases. J. Biol. Chem. 41(2013), 81-86. https://hjbc.hacettepe.edu.tr/journal/volume-41/issue-1

[70] P. P. Shah, P. R. Desai, A. R. Patel, et al. Skin permeating nanogel for the cutaneous co-delivery of two anti-inflammatory drugs. Biomaterials 33(2012), 1607-17. N. doi:10.1016/j.biomaterials.2011.11.011

[71] Kamaly, G. Fredman, M. Subramanian, S. Gadde, A. Pesic, and L. Cheung, Development and in vivo efficacy of targeted polymeric inflammation resolving nanoparticles. Proc. Natl. Acad. Sci. U.S.A. 110(2013), 6506–6511. doi:10.1073/pnas.1303377110

[72] S. Gadde, O. Even-Or, N. Kamaly, A. Hasija, P. G. Gagnon, K. H. Adusumilli, et al. Development of therapeutic polymeric nanoparticles for the resolution of inflammation. Adv. Healthc. Mater. 3(2014), 1448–1456. doi:10.1002/adhm.201300688

[73] E.J. Jeong, M. Choi, J. Lee, et al. The spacer arm length in cell-penetrating peptides influences chitosan/siRNA nanoparticle delivery for pulmonary inflammation treatment. Nanoscale, 7(2015), 20095-104, doi:10.1039/C5NR06903C

[74] M. Kumar, X. Kong, A.K. Behera, et al. Chitosan IFN-γ-pDNA Nanoparticle (CIN) Therapy for Allergic Asthma. Genet Vaccines Ther. 1(2003), 1-10. doi:10.1186/1479-0556-1-3

[75] H.-D. Lu, H.-Q. Zhao, K. Wang, et al. Novel hyaluronic acid–chitosan nanoparticles as non-viral gene delivery vector targeting osteoarthritis. Int. J. Pharm. 420 (2011), 358– 65. doi:10.1016/j.ijpharm.2011.08.046

[76] H. k.Lu, Y. Dai, L. Lulu, et al. Chitosan-Graft-Polyethylenimine/DNA Nanoparticles as Novel Non-Viral Gene Delivery Vectors Targeting Osteoarthritis. PLoS One 9(2014), 1-12. doi:10.1371/journal.pone.0084703

[77] V. Martin, I. A.C. Ribeiro, M. M. Alves, L. Gonçalves, A. J. Almeida, L. Grenho, M. H. Fernandes, C. F. Santos, P. S. Gomes, A. F. Bettencourt. Understanding intracellular trafficking and anti-inflammatory effects of minocycline chitosan-nanoparticles in human gingival fibroblasts for periodontal disease treatment. Int. J. Pharm. 572(2019), 118821. doi:10.1016/j.ijpharm.2019.118821

[78] Y. Ding, Y. Qiao, M. Wang, et al. Enhanced Neuroprotection of Acetyl-11-Keto-β-Boswellic Acid (AKBA)-Loaded O-Carboxymethyl Chitosan Nanoparticles Through Antioxidant and Anti-Inflammatory Pathways. Mol. Neurobiol. 53(2016), 3842–3853. doi:10.1007/s12035-015-9333-9

[79] A. R. Clementino, C. Marchi, M. Pozzoli, F. Bernini, F. Zimetti, F. Sonvico1, Anti-Inflammatory Properties of Statin-Loaded Biodegradable Lecithin/Chitosan Nanoparticles: A Step Toward Nose-to-Brain Treatment of Neurodegenerative Diseases, Front. Pharmacol. 24 (2021), 1-12. doi:10.3389/fphar.2021.716380

[80] B. Xiao, P. Ma, E. Viennois, D. Merlin, Urocanic acid-modified chitosan nanoparticles can confer anti-inflammatory effect by delivering CD98 siRNA to macrophages. Colloids and Surfaces B: Biointerfaces, 143 (2016), 186-193, doi:10.1016/j.colsurfb.2016.03.035

[81] K. A. Howard, S. R. Paludan, M.A. Behlke, F. Besenbacher, B. Deleuran, J. Kjems, Chitosan/siRNA Nanoparticle–mediated TNF-α Knockdown in Peritoneal Macrophages for Anti-inflammatory Treatment in a Murine Arthritis Model. Mol Ther. 17(2009), 162-8. doi: 10.1038/mt.2008.220

[82] H. M. Asif, F. Zafar, K. Ahmad, A. Iqbal, G. Shaheen, K. A. Ansari, S. Rana, R. Zahid, S. Ghaffar, Synthesis, characterization and evaluation of anti-arthritic and anti-inflammatory potential of curcumin loaded chitosan nanoparticles, Scientific Reports, 13, Article number: 10274 (2023) doi:10.1038/s41598-023-37152-7

[83] K. M. Alshehri, E. M. Abdella, Development of ternary nanoformulation comprising bee pollen-thymol oil extracts and chitosan nanoparticles for anti-inflammatory and anticancer applications, Int. J. Biol. Macromol. 242 (2023), 124584. doi:10.1016/j.ijbiomac.2023.124584

[84] A. J. Friedman, J. Phan, D. O. Schairer, J. Champer, M. Qin, A. Pirouz, K. Blecher-Paz, A. Oren, P. T. Liu, R. L. Modlin, J. Kim. Antimicrobial and Anti-Inflammatory Activity of Chitosan–Alginate Nanoparticles: A Targeted Therapy for Cutaneous Pathogens. J. Invest. Dermatol. 133 (2013), 1231-1239, doi:10.1038/jid.2012.399

[85] M. Bernela, M. Ahuja, R. Thakur. Enhancement of anti-inflammatory activity of glycyrrhizic acid by encapsulation in chitosan-katira gum nanoparticles. Eur. J. Pharm. Biopharm. 105 (2016), 141-147. doi:10.1016/j.ejpb.2016.06.003

[86] J. Tu, Y. Xu, J. Xu, Y. Ling, Y. Cai. Chitosan nanoparticles reduce LPS-induced inflammatory reaction via inhibition of NF-κB pathway in Caco-2 cells. Int. J. Biol. Macromol. 86(2016), 848-856. doi:10.1016/j.ijbiomac.2016.02.015

[87] S. Muzammil, J. Neves Cruz, R. Mumtaz, I. Rasul, S. Hayat, M.A. Khan, A.M. Khan, M.U. Ijaz, R.R. Lima, M. Zubair, Effects of Drying Temperature and Solvents on In Vitro Diabetic Wound Healing Potential of Moringa oleifera Leaf Extracts, Molecules. 28 (2023). doi:10.3390/molecules28020710

[88] N. Jabbari, Z. Eftekhari, N. H. Roodbari, K. Parivar. Evaluation of Encapsulated Eugenol by Chitosan Nanoparticles on the aggressive model of rheumatoid arthritis, Int. Immunopharmacol. 85(2020), 106554, doi: 10.1016/j.intimp.2020.106554

[89] A. M. Abd El-Hameed. Polydatin-loaded chitosan nanoparticles ameliorates early diabetic nephropathy by attenuating oxidative stress and inflammatory responses in streptozotocin-induced diabetic rat. J. Diabetes Metab. Disord. 19(2020), 1599–1607. doi:10.1007/s40200-020-00699-7

[90] C. Théry, M. Boussac, P. Véron, P. Ricciardi-Castagnoli, G. Raposo, J. Garin, S. Amigorena, Proteomic analysis of dendritic cell-derived exosomes: a secreted subcellular compartment distinct from apoptotic vesicles. Journal of Immunology (Baltimore, Md.: 1950), 166(2002), 7309-7318. doi:10.4049/jimmunol.166.12.7309

[91] J. Wang, G. Chen, H. Jiang, Z. Li, and X.Wang, Advances in nano-scaled biosensors for biomedical applications. The Analyst, 138 (2013), 4427-4435. https://doi.org/10.1039/C3AN00438D

[92] E. I. Buzas, B. Gyorgy, G. Nagy, A. Falus, S. Gay, Emerging role of extracellular vesicles in inflammatory diseases. Nat. Rev. Rheumatol. 10 (2014), 356-364. DOI: 10.1038/nrrheum.2014.19

[93] I.N. de F. Ramos, M.F. da Silva, J.M.S. Lopes, J.N. Cruz, F.S. Alves, J. de A.R. do Rego, M.L. da Costa, P.P. de Assumpção, D. do S. Barros Brasil, A.S. Khayat, Extraction, Characterization, and Evaluation of the Cytotoxic Activity of Piperine in Its Isolated form and in Combination with Chemotherapeutics against Gastric Cancer, Molecules. 28 (2023). DOI: 10.3390/molecules28145587

[94] H.B. Koh, H.J. Kim, S.W. Kang, T.H. Yoo. Exosome-Based Drug Delivery: Translation from Bench to Clinic. Pharmaceutics,15(2023):2042. doi: 10.3390/pharmaceutics15082042.

[95] X. Zhuang, X. Xiang, W. Grizzle, L. Steinman, D. Miller, H.-G. Zhang, et al. Treatment of Brain Inflammatory Diseases by Delivering Exosome Encapsulated Anti-inflammatory Drugs from the Nasal Region to the Brain. Mol. Ther. 19 (2011), P1769-1779. DOI: 10.1038/mt.2011.164

[96] C. Liu, X. Yan, Y. Zhang, et al. Oral administration of turmeric-derived exosome-like nanovesicles with anti-inflammatory and pro-resolving bioactions for murine colitis therapy. J. Nanobiotechnol 20(2022), 206. https://doi.org/10.1186/s12951-022-01421-w

[97] Q. Lin, M. Qu, B. Zhou, H. K. Patra, Z, Sun, Q, Luo, W, Yang, Y, Wu, Y, Zhang, L, Li, L, Deng, L, Wang, T, Gong, Q, He, L, Zhang, X, Sun, Z, Zhang. Exosome-like nanoplatform modified with targeting ligand improves anti-cancer and anti-inflammation effects of imperialine. J. Control. Release 311–312 (2019), 104-116. DOI: 10.1016/j.jconrel.2019.08.037

[98] Z. B. Deng, Y. Liu, C. Liu, X. Xiang, J. Wang, Z. Cheng, et al. Immature myeloid cells induced by a high-fat diet contribute to liver inflammation. Hepatology 50(2009), 1412–1420. DOI: 10.1002/hep.23148

[99] D. Sun, X. Zhuang, X. Xiang, Y. Liu, S. Zhang, C. Liu, et al. A novel nanoparticle drug delivery system: the anti-inflammatory activity of curcumin is enhanced when encapsulated in exosomes. Mol. Ther. 18(2010), 1606–1614. DOI: 10.1038/mt.2010.105

[100] C. Yang, and D. Merlin, Nanoparticle-mediated drug delivery systems for the treatment of IBD: current perspectives. Int. J. Nanomed. 14(2019), 8875–8889. DOI: 10.2147/IJN.S210315

[101] Z. Cai, W. Zhang, F. Yang, L. Yu, Z. Yu, J. Pan, et al. Immunosuppressive exosomes from TGF-b1 gene-modified dendritic cells attenuate Th17-mediated inflammatory

autoimmune disease by inducing regulatory T cells. Cell Res. 22(2012), 607–610. DOI: 10.1038/cr.2011.196

[102] Y. Wang, J. Tian, X. Tang, K. Rui, X. Tian, J. Ma, et al. Exosomes released by granulocytic myeloid-derived suppressor cells attenuate DSS induced colitis in mice. Oncotarget 7 (2016), 15356–15368. DOI: 10.18632/oncotarget.7324

[103] G. Wu, J. Zhang, Q. Zhao, W. Zhuang, J. Ding, C. Zhang, et al. Molecularly engineered macrophage-derived exosomes with inflammation tropism and intrinsic heme biosynthesis for atherosclerosis treatment. Angew. Chem. Int. Ed. Engl. 59 (2020), 4068–4074. DOI: 10.1002/anie.201913700

[104] F. Yan, Z. Zhong, Y. Wang, Y. Feng, Z. Mei, H. Li, et al. Exosome-based biomimetic nanoparticles targeted to inflamed joints for enhanced treatment of rheumatoid arthritis. J. Nanobiotechnol. 18 (2020), 115. DOI: 10.1186/s12951-020-00675-6

[105] K. Asadullah, W. Sterry, and H. D. Volk, Interleukin-10 therapy–review of a new approach. Pharmacol. Rev. 55 (2003), 241–269. DOI: 10.1124/pr.55.2.4

[106] H. S. Chiong, Y. K. Yong, Z. Ahmad, M. R. Sulaiman, Z. A. Zakaria, K. H. Yuen, et al. Cytoprotective and enhanced anti-inflammatory activities of liposomal piroxicam formulation in lipopolysaccharide-stimulated RAW 264.7macrophages. Int. J. Nanomed. 8 (2013), 1245–1255. doi: 10.2147/IJN.S42801

[107] R. Duivenvoorden, J. Tang, D. P. Cormode, A. J. Mieszawska, D. Izquierdo-Garcia, C. Ozcan, M. J. Otten, N. Zaidi, M. E. Lobatto, S. M. van Rijs, et al. A Statin-Loaded Reconstituted High-Density Lipoprotein Nanoparticle Inhibits Atherosclerotic Plaque Inflammation. Nat. Commun. 5(2014), 3065. DOI: 10.1038/ncomms4065

[108] M. E. Lobatto, Z. A. Fayad, S. Silvera, E. Vucic, C. Calcagno, V. Mani, S. D. Dickson, K. Nicolay,; M. Banciu, R. M. Schiffelers et al. Multimodal Clinical Imaging to Longitudinally Assess a Nanomedical Anti-Inflammatory Treatment in Experimental Atherosclerosis. Mol. Pharm. 7(2010), 2020–2029. DOI: 10.1021/mp100309y

[109] F. M. van der Valk, D. F. van Wijk, M. E. Lobatto, H. J. Verberne, G. Storm, M. C.M. Willems, D. A. Legemate, A. J. Nederveen, C. Calcagno, V. Mani, S. Ramachandran, et. al. Prednisolone-Containing Liposomes Accumulate in Human Atherosclerotic Macrophages upon Intravenous Administration. Nanomedicine 11(2015), 1039–1046. DOI: 10.1016/j.nano.2015.02.021

[110] K. Takeda, B. E. Clausen, T. Kaisho, T. Tsujimura, N. Terada, I. Förster, S. Akira. Enhanced Th1 activity and development of chronic enterocolitis in mice devoid of Stat3 in macrophages and neutrophils. Immunity 10(1999), 39–49. DOI: 10.1016/s1074-7613(00)80005-9

[111] R. Toita, T. Kawano, M. Murata, J. H. Kang, Anti-obesity and anti-inflammatory effects of macrophage-targeted interleukin-10-conjugated liposomes in obese mice. Biomaterials 110(2016), 81–88. doi:10.1016/j.biomaterials.2016.09.018

[112] S. Nagata, R. Hanayama, and K. Kawane, Autoimmunity and the clearance of dead cells. Cell 140(2010), 619–630. 10.1016/j.cell.2010.02.014

[113]   Y. Wu, M. Sun, D. Wang, G. Li, J. Huang, S. Tan, et al. A PepT1mediated medicinal nano-system for targeted delivery of cyclosporine A to alleviate acute severe ulcerative colitis. Biomater. Sci. 7(2019a), 4299–4309. DOI: 10.1039/C9BM00925F

[114]   Y. Wu, Y. Zhang, L. Dai, Q. Q. Wang, L. J. Xue, Z. Su, et al. An apoptotic body-biomimic liposome in situ upregulates anti-inflammatory macrophages for stabilization of atherosclerotic plaques. J. Control. Release 316(2019b), 236–249. DOI: 10.1016/j.jconrel.2019.10.043

[115]   X. Xu, Y. Li, L. Wang, Y. Li, J. Pan, X. Fu, et al. Triple functional polyether ether ketone surface with enhanced bacteriostasis and anti-inflammatory and osseo integrative properties for implant application. Biomaterials 212 (2019), 98–114.DOI: 10.1016/j.biomaterials.2019.05.014

[116]   V. Bagalkot, J. A. Deiuliis, S. Rajagopalan, and A. Maiseyeu, "Eat me" imaging and therapy. Adv. Drug Deliv. Rev. 99(2016), 2–11. doi: 10.1016/j.addr.2016.01.009

[117]   Oh, J.; Drumright, R. J.; Siegwart, D.; Matyjaszewski, K. The development of microgels/nanogels for drug delivery applications. Prog. Polym. Sci. 2008, 33, 448– 477, doi:10.1016/j.progpolymsci.2008.01.002

[118]   M. Kaur, K. Sudhakar, V. Mishra. Fabrication and biomedical potential of nanogels: An overview. Int. J. Polym. Mater. Polym. Biomater. 2018, 287-296. DOI:10.1080/00914037.2018.1445629

[119]   W. Wang, L. Sun, P. Zhang, J. Song, W. Liu, An anti-inflammatory cell-free collagen/resveratrol scaffold for repairing osteochondral defects in rabbits. Acta Biomater. 10(2014), 4983–4995. DOI: 10.1016/j.actbio.2014.08.022

[120]   J. Ratanavaraporn, H. Furuya, Y. Tabata, Local suppression of proinflammatory cytokines and the effects in BMP-2-induced bone regeneration. Biomaterials 33 (2012), 304–316. DOI: 10.1016/j.biomaterials.2011.09.050

[121]   R. M. Gower, R. M. Boehler, S. M. Azarin, C. F. Ricci, J. N. Leonard, L. D. Shea, Modulation of leukocyte infiltration and phenotype in microporous tissue engineering scaffolds via vector induced IL-10 expression. Biomaterials 35(2014), 2024–2031. doi:10.1016/j.biomaterials.2013.11.036

[122]   K. Nakamura, S. Yokohama, M. Yoneda, S. Okamoto, Y. Tamaki, T. Ito, et al. High, but not low, molecular weight hyaluronan prevents T-cell-mediated liver injury by reducing proinflammatory cytokines in mice. J. Gastroenterol. 39(2004), 346–354. DOI: 10.1007/s00535-003-1301-x

[123]   A. N. Kuskov, P. P. Kulikov, A. V. Goryachaya, M. N. Tzatzarakis, A. O. Docea, K. Velonia, et al. Amphiphilic poly-N-vinylpyrrolidone nanoparticles as carriers for non-steroidal, anti-inflammatory drugs: in vitro cytotoxicity and in vivo acute toxicity study. Nanomedicine 13(2017), 1021–1030. https://doi.org/10.3390/pharmaceutics14050925

[124]   Y. Wang, A thrombin-triggered self-regulating anticoagulant strategy combined with anti-inflammatory capacity for blood-contacting implants. Sci. Adv. 8(2022), eabm3378. DOI: 10.1126/sciadv.abm3378

[125] N. Aminu, S.-Y. Chan, M.-F. Yam, S.-M. Toh. A dual-action chitosan-based nanogel system of triclosan and flurbiprofen for localised treatment of periodontitis. Int. J. Pharm. 570 (2019), 118659. DOI: 10.1016/j.ijpharm.2019.118659

[126] A. Goel, F. J. Ahmad, R. M. Singh, G. N. Singh Anti-inflammatory activity of nanogel formulation of 3-acetyl11-keto-β-boswellic acid, Pharmacologyonline 3(2009), 311-318. https://api.semanticscholar.org/CorpusID:59496153

[127] G. S. L Singka, N. A. Samah, M. H. Zulfakar, A. Yurdasiper, C. M. Heard. Enhanced topical delivery and anti-inflammatory activity of methotrexate from an activated nanogel. Eur. J. Pharm. Biopharm. 76 (2010), 275-281. DOI: 10.1016/j.ejpb.2010.06.014

[128] Y.-F. Chen, G.-Y. Chen, C.-H. Chang, Y.-C. Su, Y.-C. Chen, Y. Jiang, J.-S. Jan. TRAIL encapsulated to polypeptide-crosslinked nanogel exhibits increased anti-inflammatory activities in Klebsiella pneumoniae-induced sepsis treatment. Materials Science and Engineering: C, 102 (2019), 85-95. DOI: 10.1016/j.msec.2019.04.023

[129] T. Li, J. Yang, C. Weng, P. Liu, Y. Huang, S. Meng, R. Li, L. Yang, C. Chen, X. Gong. Intra-articular injection of anti-inflammatory peptide-loaded glycol chitosan/fucoidan nanogels to inhibit inflammation and attenuate osteoarthritis progression. Int. J. Biol. Macromol. 170 (2021), 469-478. DOI: 10.1016/j.ijbiomac.2020.12.158

[130] C. Valentino, B. Vigani, I. Fedeli, D. Miele, G. Marrubini, L. Malavasi, F. Ferrari, G. Sandri, S. Rossi. Development of alginate-spermidine micro/nanogels as potential antioxidant and anti-inflammatory tool in peripheral nerve injuries. Formulation studies and physico-chemical characterization. Int. J. Pharm. 626 (2022), 122168. DOI: 10.1016/j.ijpharm.2022.122168

[131] S. Obuobi, K. Julin, E.G.A. Fredheim, Mona Johannessen, N.Škalko-Basnet. Liposomal delivery of antibiotic loaded nucleic acid nanogels with enhanced drug loading and synergistic anti-inflammatory activity against S. aureus intracellular infections. Journal of Controlled Release 324 (2020), 620-632. DOI: 10.1016/j.jconrel.2020.06.002

[132] J. Yeo, J. Lee, S. Yoon and W. J. Kim, Tannic acid-based nanogel as an efficient anti-inflammatory agent. Biomater Sci 30(2013), 257–268. DOI: 10.1039/c9bm01384a

[133] P. P. Shah, P. R. Desai, A. R. Patel, M. S. Singh. Skin permeating nanogel for the cutaneous co-delivery of two anti-inflammatory drugs. Biomaterials 33 (2012), 1607-1617. DOI: 10.1016/j.biomaterials.2011.11.011

[134] S. M. Gutowski, J. T. Shoemaker, K. L. Templeman, Y. Wei, R. A. Latour, R. V. Bellamkonda, et al. Protease-degradable PEG-maleimide coating with on-demand release of IL-1Ra to improve tissue response to neural electrodes. Biomaterials 44(2015), 55–70. DOI: 10.1016/j.biomaterials.2014.12.009

[135] M. Giulbudagian, G. Yealland, S. Hönzke, A. Edlich, B. Geisendörfer, B. Kleuser, S. Hedtrich, and M. Calderón. Breaking the Barrier - Potent Anti-Inflammatory Activity following Efficient Topical Delivery of Etanercept using Thermoresponsive Nanogels. Theranostics 8(2018), 450–463. DOI: 10.7150/thno.21668

[136] F. Esmaeili, M. Zahmatkeshan, Y. Yousefpoor, H. Alipanah, E. Safari, M. Osanloo. Anti-inflammatory and anti-nociceptive effects of Cinnamon and Clove essential oils nanogels:

Materials Research Forum LLC
https://doi.org/10.21741/9781644903858-6

an in vivo study. BMC Complement. Med. Ther. 22, 143 (2022). DOI: 10.1186/s12906-022-03619-9

[137] P.-H. Lin, H.-J. Jian, Y.-J. Li, Y.-F. Huang, A. Anand, C.-C. Huang, H.-J. Lin, J.-Y. Lai. Alleviation of dry eye syndrome with one dose of antioxidant, anti-inflammatory, and mucoadhesive lysine-carbonized nanogels, Acta Biomaterialia, 141 (2022), 140-150. Doi: 10.1016/j.actbio.2022.01.044.

[138] R. Gul, N. Ahmed, N. Ullah, M. I. Khan, A. Elaissari, and U. A. Rehman, Biodegradable ingredient-based emulgel loaded with ketoprofen nanoparticles. AAPS Pharm. Sci. Tech. 19(2018), 1869–1881. DOI: 10.1208/s12249-018-0997-0

[139] R. E. Whitmire, D. S. Wilson, A. Singh, M.E. Levenston, N. Murthy, A. J. García, Self-assembling nanoparticles for intra-articular delivery of anti-inflammatory proteins. Biomaterials 33(2012),7665–7675. DOI: 10.1016/j.biomaterials.2012.06.101

[140] M. J. Webber, J. B. Matson, V. K. Tamboli, and S. I. Stupp, Controlled release of dexamethasone from peptide nanofiber gels to modulate inflammatory response. Biomaterials 33(2012), 6823–6832. DOI: 10.1016/j.biomaterials.2012.06.003

[141] F. Leuschner, P. Dutta, R. Gorbatov, T. I. Novobrantseva, J. S. Donahoe, G. Courties, et al. Therapeutic siRNA silencing in inflammatory monocytes in mice. Nat. Biotechnol. 29(2011), 1005–1010. DOI: 10.1038/nbt.1989

[142] S. Park, S. Kang, X. Chen, E. J. Kim, J. Kim, N. Kim, et al. Tumor suppression via paclitaxel-loaded drug carriers that target inflammation marker upregulated in tumor vasculature and macrophages. Biomaterials 34(2013), 598–605. DOI: 10.1016/j.biomaterials.2012.10.004

[143] H. Agarwal, A. Nakara, V. K. Shanmugam, Anti-inflammatory mechanism of various metal and metal oxide nanoparticles synthesized using plant extracts: A review, Biomedicine and Pharmacotherapy, 109 (2019), 2561-2572, DOI: 10.1016/j.biopha.2018.11.116

[144] S. Belperain, Z.Y. Kang, A. Dunphy, B. Priebe, NHL Chiu, Z. Jia, Anti-Inflammatory Effect and Cellular Uptake Mechanism of Carbon Nanodots in Human Microvascular Endothelial Cells. Nanomaterials (Basel). 11(2021), 1247. DOI: 10.3390/nano11051247

[145] M. Summer, R. Ashraf, S. Ali, H. Bach, S. Noor, Q. Noor, S. Riaz, R. R. M. Khan, Inflammatory response of nanoparticles: Mechanisms, consequences, and strategies for mitigation. Chemosphere, 363(2024), 142826, https://doi.org/10.1016/j.chemosphere.2024.142826.

Nanomaterials in Biological Systems
Materials Research Foundations 185 (2026) 154-163

Materials Research Forum LLC
https://doi.org/10.21741/9781644903858-7

Chapter 7

Environmental Impacts of Nanoparticles

Jyoti Chawla*, Rajeev Kumar

Manav Rachna International Institute of Research and Studies, Faridabad Haryana India

*jyoti.chawla@gmail.com

Abstract

Nanomaterials due to their unique properties offered many new opportunities for technological developments in different industries. Diverse ranges of nanomaterials have been developed for different applications and further tailoring of their structure and surface area for desirable utility is a priority area of researchers. These materials are very promising and have become part of everyday life. Nano materials are chemically and biologically active due to high surface area and due to their increasing demand, production and utilization concerns have been raised about the fate in the environment as well as their impact on environment quality. Nevertheless, nanomaterials also offer great potential for environmental remediation. In the present chapter, environmental impact of nanomaterials has been discussed and related challenges have been analyzed in terms of monitoring, risk assessment and fate.

Keywords

Nanomaterials; Environment; Carbon Nanotubes; Remediation; Risk Assessment

Contents

1.   Introduction

Nanomaterials due to their unique attributes are used in numerous ways in almost every sector including research, agriculture, health care, construction, food industry, textile industry, electronics etc. Nanoparticles like carbon-based nanomaterials, metal-based nanomaterials, nano composites contributed in remarkable development of conventional technologies. Due to their increasing demand, production and utilization concerns have been raised about the fate in the environment as well as their impact on environment quality. Nanomaterials also occur in nature and can also be released during volcanic eruptions, weathering of minerals, forest fires, dust storms etc. and these particles also interact with components of environment [1]. Integrated approach for risk assessment of specific nanoparticles considering their origin and all possible sources is required to understand the impact of nanoparticles on environment.

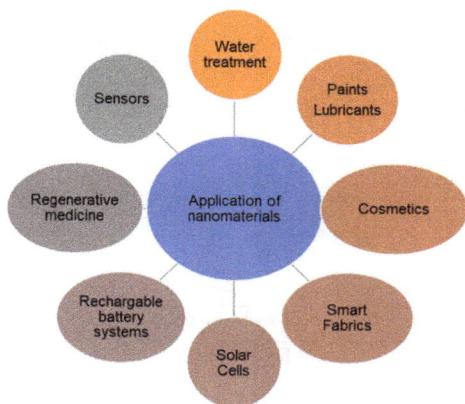

Figure 1: Applications of nanomaterials in different sectors

Various studies have been conducted for environmental and human health risk assessment of the nanomaterials. Numerous strategies for safe utilization of nanomaterials have also been suggested. Many researchers have discussed the fate of specific nanomaterials in the environment [2,3] but still there are many knowledge gaps due to introduction of many new materials on regular basis. Concentrations of some nanoparticles in different environmental samples have also been monitored/predicted in few studies [4]. Toxicity studies of different nanomaterials including silver nano particles, fullerenes, nano titanium oxide, carbon nanotubes etc. further help to determine the environmental and health hazards. These studies are very important as they predict the unintended hazards of numerous synthesized nanomaterials for technological development. There are still many challenges related to the potential environmental implications of nanoparticles because of their increasing use in multiple sectors (Figure 1). Many materials have been tailored for these applications including functionalized carbon nano tubes, fullerenes, graphenes, quantum dots, dendrimers, cerium oxide, zinc oxide, titanium oxide etc. Each material due to its unique characteristics may pose different environmental hazards in different concentrations. Risk assessment of nanoparticles in general is complex due to change in their aggregation behavior and stability with response to change in surface area, charge, shape and elemental composition. These variations in the characteristics of nanoparticles makes difficult to predict the generalized environmental and health hazards. So, it is important to identify the prominent nanoparticles that have tendency to accumulate in environment and conduct the separate risk assessment studies. In the present chapter, environmental impact of commonly used nanomaterials have been discussed and related challenges have been analysed in terms of monitoring, risk assessment and fate.

2. Monitoring/Prediction studies of nanomaterials in environment

There are only few monitoring studies of specific nanomaterials in environment due to many reasons including appropriate sample collection, preservation and analysis. However, some researchers estimated/predicted environmental concentrations of nanomaterials in water, air and soil samples through modelling studies (Table 1).

Table 1 Predicted environmental concentrations of nanomaterials in water

Type of Nanomaterial	Predicted Concentration in surface water	Predicted Concentration in sludge/soil	Reference
Fullerene	0.003 ng/l	1 ng/kg	[5]
Fullerene	-	44.7 µg/kg	[6] Batley and Mclaughlin 2010
TiO$_2$	-	1030 µg/kg	[6] Batley and Mclaughlin 2010
TiO$_2$	21 ng/l	89 g/kg	[5]
TiO$_2$	-	3 ng/l - 1.6 mg/l	[7]
TiO$_2$	2.17µg /l	Freshwater	[8] Sun et al. 2016
TiO$_2$	-	0.4 µg/kg	[4]
nano-silver	-	0.02µg/kg	[4]
nano-silver	-	1.45 µg/kg	[6] Batley and Mclaughlin 2010
nano-silver	-	4.3 -13.03 µg/kg	[9] Hendren et al. in 2013
ZnO	-	3190 µg/kg	[6] Batley and Mclaughlin 2010
ZnO	0.38µg /l	-	[8] Sun et al. 2016
Carbon nanotubes	-	0.01µg/kg	[4]

Carbon based nanomaterials including fullerene, carbon nano tubes and metal-based nanomaterials including ZnO, TiO2, nano silver are most commonly reported nanomaterials in environmental matrices.

3. Impact of nanoparticles in environment

Various studies have been conducted to assess the environmental and health impact of nanomaterials. It has been confirmed that nanoparticle exposure increases oxidative stress and cytotoxicity in studied target organisms. It has also been found that the nanoparticles have tendency to accumulate in soil and affects the growth of plants growing in soil contaminated with metal-based nanoparticles. Also, the positive aspects of nanomaterials also cannot be ignored as these may be utilized for remediation of contaminants from environment. Water treatment and

Nanomaterials in Biological Systems   Materials Research Forum LLC
Materials Research Foundations 185 (2026) 154-163   https://doi.org/10.21741/9781644903858-7

air/soil quality improvement are few promising applications of nanomaterials [10-13] like graphene, metal oxides have been applied as adsorbents for remediation of contaminants.

3.1   Ecological Impacts of carbon based nanomaterials

Carbon based nanomaterials including pristine and functionalized fullerenes, carbon nanotubes, graphenes, nano-diamond have promising applications in various fields due to their high conductivity, ability to withstand high temperatures and pressures, excellent mechanical strength, anti-oxidative properties and optical behaviour. The fate and behaviour of these materials have been explained in different studies [14,15]. It has been observed that carbon-based nanomaterials with different surface area/functionalization/geometric structure show relatively different toxicity profiles. The trend for cytotoxicity of carbon-based nanomaterials based on various studies is found as single-walled carbon nanotubes > multiwalled carbon nanotubes > carbon black powders/fullerenes > nano diamonds [16]. Carbon based nanoparticles may enter the environment during production, utilization and disposal of related products. Due to good adsorption capacity and non-degradable nature these materials may combine with other substances/ accumulate in the environment and pose adverse effects on organisms [17,18]. Chung et al., 2011 during a study to observe the impact of multi-walled carbon nanotubes on soil microbes confirmed that multi-walled carbon nanotubes can inhibit the biomass and activity of microorganisms in the soil when present in high concentration [19]. It was also highlighted in another study by Dharni et al., 2016 that carbon nanotubes inhibit the growth of fungi and bacteria [20]. Kang et al., 2019 also confirmed that carbon nanotubes reduce the content of microbial carbon in soil and affect the soil microbes [21]. Gamoń et al., 2023 used Lemnaminor, Artemiafranciscana, Brachionuscalyciflorus, Thamnocephalusplatyuru and Escherichia coli to estimate the ecotoxic effect of carbon nanomaterials graphene oxide, reduced graphene oxide, pristine and fumctionalized multi-walled carbon nanotubes and it was observed that the studied materials can be toxic at all trophic levels in a short exposure time and extent of toxicity varies with material dose and structure and graphene oxide was found to safest among all [22]. Germination of tomato seedlings was studied by Khodakovskaya et al., 2011 in presence of graphene and carbon nanotubes and during investigation of genotoxicity of materials on leaves/roots results confirmed that carbon nanotubes induce more activation stress [23].

Figure 2 represents the impact of carbon-based nanomaterials on environment highlighting the results of some important studies. In addition to these effects, the applications of carbon-based nanomaterials in remediation/monitoring of pollutants cannot be ignored as they may also be employed to improve the air/water/soil quality (Figure 3). Carbon nanotubes and graphenes can be used for removal heavy metal ions from water and soil. These materials may also be utilized to improve ambient air quality and reduce noise pollution. Yang et al. 2013 used silica nanoparticles to remove lead present in air [25]. During the risk assessment studies all the aspects are required to be considered. However, there are many knowledge gaps in the area due to certain challenges like the clarity on types of carbon-based materials that enter the environment, their concentration and the experimental conditions maintained during toxicity studies. Robust tools are required for estimation of different carbon-based nanomaterials present in environment in addition to ecotoxicity assays for different forms of carbon-based nanomaterials to address the challenges and knowledge gaps.

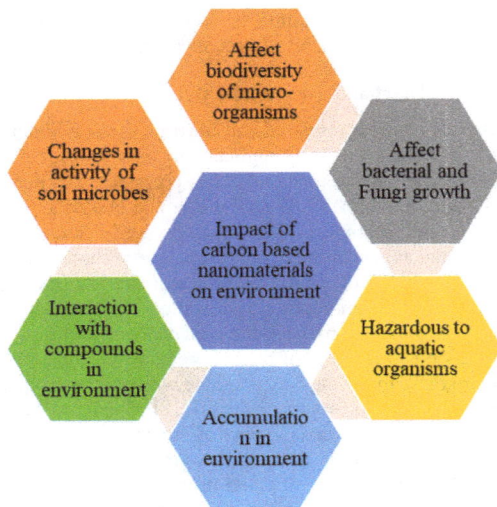

Figure 2: Impact of carbon-based nanomaterials on environment

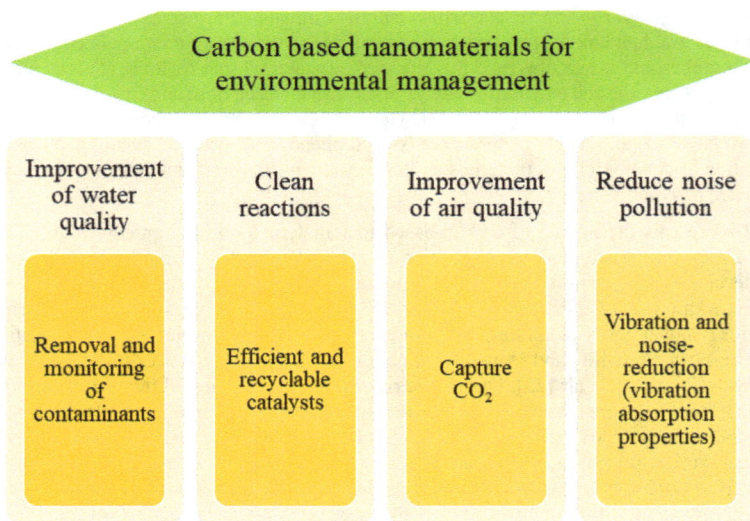

Figure 3: Applications of carbon based nanomaterials to enhance environment quality

3.2 Ecological Impacts of metal based nanomaterials

There are innumerable applications of metal based nanomaterials in different fields and some applications are quite useful for tackling environmental issues. However, there are also concerns due to their adverse effects on environment and human health during their synthesis and disposal [26]. Release of metallic NPs into the environmental systems from different sources may pose deleterious impact on atmosphere, microorganisms, plants etc. Release of nanoparticles and their persistence in soil decimate its beneficial microbial composition and disturb the beneficial interactions between soil and plant-microbes. Feizi et al., 2013 in a study confirmed that TiO_2 nanoparticles with concentration more than 40ppm hamper the seed germination, and growth of fennel [27]. It was also observed during study of effect of TiO_2 nanoparticles on maize that nanoparticles reduced the chlorophyll content as well as seedling growth, but increase in uptake of phosphorous, nitrogen and potassium from the soil was also reported [28]. In another study (Mattiello et al., 2015) to assess the effect of CeO_2 nano materials and TiO_2 nanomaterials on barley, no penetration of nanoparticles was observed in hard seed coats but genotoxicity and phototoxicity was confirmed in the root cells [29].

Metal based nanomaterials may also be used as insecticides, fertilizers, and pesticides. In agriculture, around 40% nanoparticles are used for fertilization around 25% for animal husbandry, around 20% for plant protection, 10% for plant breeding, and 5% for soil enhancement, according to [30]. Lots of metallic nanoparticales are possible replacements for conventional nutrient sources as slow-releasing fertilizers because they include important micronutrients for plants. As an effective, slowly-releasing zinc fertilizer, zinc oxide nanoparticles have been investigated as a means of reducing the pervasive Zn deficit in soils. As a slow-releasing supply for the long-term Cu requirements of crops, copper oxide nanoparticles have also been applied as slowly-releasing fertilizer. Silicon oxide nanoparticles have been created as a slow-releasing silicon supplement to reduce salt stress and as a transporter of important macronutrients such as N, P, and K for plants. Lots of MNPs have also been investigated as more effective insecticides and disease control agents because of their antibacterial characteristics. Copper, zinc, titanium, silicon and silver based nanoparticles or nanocomposites demonstrated antibacterial and antifungal activities against a variety of diseases associated with plants [31-33].

Interactions between MNPs and soil clay minerals have been investigated for alteration in the soil attributes by many researchers. Addition of 6% copper oxide NPs or aluminium oxide NPs to clay (kaolinite) increased shrinkage by 8–17%, which improved soil's ability to keep its aggregate structure under dry circumstances. However, clay shrinkage was not observed when the concentration of these NPs was decreased to 0.5%. Addition of 2% copper oxide NPs or aluminium oxide NPs to clay (kaolinite) decreased the hydraulic conductivity of soil by 30-45% with pore blockage. In particular for soil microorganisms, the use of metallic nanoparticles in agriculture may disrupt the environment [34]. Numerous microbial processes, including nutrient cycling, compound mineralization, and organic matter synthesis, can be adversely impacted by MNPs. Because they focus their energy on maintaining their cells rather than on microbiological functions crucial for soil health, microorganisms under stress in the presence of pollutants have low metabolic efficiency. The crucial factor that affects its operation is soil. While several MNPs can change soil pH, the effects can vary. At 1-100 mg/kg, titanium, ceria, or copper oxide nanoparticles increased loam soil's pH by 0.4 units while lowering it by 0.1-0.4 units in a sandy loam soil. There is a need of more experiments to be conducted to understand the behaviour of different types of metal based nanomaterials in real environmental conditions so that these materials may be utilized in a better way without harming the environment [35].

Conclusion

Various studies conducted to assess the environmental and health impact of nanomaterials confirmed that nanoparticles have tendency to accumulate in environment and pose adverse effects on ecosystem. Presence of the nanoparticles in environment could lead to imbalance of microbial and environmental processes and cause many health issues. Integrated approach considering all the pathways for specific nanomaterial in environment along with reliable monitoring tools is required for risk assessment of different nanomaterials throughout their lifecycle in different environmental matrices. Studies are also required to assess the effects of multiple nanomaterials when present together in the environment to predict their potential risks. Applications of nanomaterials in environment protection also cannot be ignored as these materials may be utilized for development of effective remediation technology. So, there is a need of extensive research focussed on sensing of nanoparticles in environment followed by monitoring of concentration of concerned nanoparticles and related risk assessment in real scenario conditions. Accordingly, the strategies for safer use of nanomaterials may be proposed in order to minimise their adverse effects on environment and health. Experiments may be planned in phased manner giving priority to the leading materials as per their production and presence in environment.

References

[1] S. Wagner, A. Gondikas, E. Neubauer, T. Hofmann, F. Von Der Kammer, Spot the difference: Engineered and natural nanoparticles in the environment-release, behavior, and fate, Angew. Chemie - Int. Ed. 53 (2014) 12398–12419. https://doi.org/10.1002/anie.201405050

[2] M. Farré, J. Sanchís, D. Barceló, Analysis and assessment of the occurrence, the fate and the behavior of nanomaterials in the environment, TrAC - Trends Anal. Chem. 30 (2011) 517–527. https://doi.org/10.1016/j.trac.2010.11.014

[3] D. Lin, X. Tian, F. Wu, B. Xing, Fate and Transport of Engineered Nanomaterials in the Environment, J. Environ. Qual. 39 (2010) 1896–1908. https://doi.org/10.2134/jeq2009.0423.

[4] N.C. Mueller, B. Nowack, Exposure modeling of engineered nanoparticles in the environment, Environ. Sci. Technol. 42 (2008) 4447–4453. https://doi.org/10.1021/es7029637

[5] G.E. Batley, J.K. Kirby, M.J. McLaughlin, Fate and risks of nanomaterials in aquatic and terrestrial environments, Acc. Chem. Res. 46 (2013) 854–862. https://doi.org/10.1021/ar2003368

[6] F. Gottschalk, C. Lassen, J. Kjoelholt, F. Christensen, B. Nowack, Modeling flows and concentrations of nine engineered nanomaterials in the Danish environment, Int. J. Environ. Res. Public Health. 12 (2015) 5581–5602. https://doi.org/10.3390/ijerph120505581

[7] T.Y. Sun, N.A. Bornhöft, K. Hungerbühler, B. Nowack, Dynamic Probabilistic Modeling of Environmental Emissions of Engineered Nanomaterials, Environ. Sci. Technol. 50 (2016) 4701–4711. https://doi.org/10.1021/acs.est.5b05828

[8]C.O. Hendren, M. Lowry, K.D. Grieger, E.S. Money, J.M. Johnston, M.R. Wiesner, S.M. Beaulieu, Modeling approaches for characterizing and evaluating environmental exposure to engineered nanomaterials in support of risk-based decision making, Environ. Sci. Technol. 47 (2013) 1190–1205. https://doi.org/10.1021/es302749u

[9]R. Kumar, J. Chawla, Removal of Cadmium Ion from Water/Wastewater by Nano-metal Oxides: A Review, Water Qual. Expo. Heal. 5 (2014) 215–226. https://doi.org/10.1007/s12403-013-0100-8

[10]   R. Kumar, J. Chawla, Removal of Cadmium Ion from Water/Wastewater by Nano-metal Oxides: A Review, Water Qual. Expo. Heal. 5 (2014) 215–226. https://doi.org/10.1007/s12403-013-0100-8

[11]   J. Chawla, R. Kumar, I. Kaur, Carbon nanotubes and graphenes as adsorbents for adsorption of lead ions from water: A review, J. Water Supply Res. Technol. - AQUA. 64 (2015) 641–659. https://doi.org/10.2166/aqua.2015.102

[12]   B. Pan, B. Pan, W. Zhang, L. Lv, Q. Zhang, S. Zheng, Development of polymeric and polymer-based hybrid adsorbents for pollutants removal from waters, Chem. Eng. J. 151 (2009) 19–29. https://doi.org/10.1016/j.cej.2009.02.036

[13]   R.J. Williams, S. Harrison, V. Keller, J. Kuenen, S. Lofts, A. Praetorius, C. Svendsen, L.C. Vermeulen, J. van Wijnen, Models for assessing engineered nanomaterial fate and behaviour in the aquatic environment, Curr. Opin. Environ. Sustain. 36 (2019) 105–115. https://doi.org/10.1016/j.cosust.2018.11.002

[14]   R. Madannejad, N. Shoaie, F. Jahanpeyma, M.H. Darvishi, M. Azimzadeh, H. Javadi, Toxicity of carbon-based nanomaterials: Reviewing recent reports in medical and biological systems, Chem. Biol. Interact. 307 (2019) 206–222. https://doi.org/10.1016/j.cbi.2019.04.036

[15]   J. Chawla, D. Singh, B. Sundaram, A. Kumar, Identifying Challenges in Assessing Risks of Exposures of Silver Nanoparticles, Expo. Heal. 10 (2018) 61–75. https://doi.org/10.1007/s12403-017-0245-y

[16]   X. Geng, Y. Guo, Y. Wang, Y. Zhai, B. Xu, L. Qu, Z. Li, Revealing the Adverse Effects of Trace Amount Broad-Spectrum Antimicrobial: a Direct and Sensitive Visualization Method Based on Carbon Nanoprobe, Adv. Funct. Mater. 32 (2022). https://doi.org/10.1002/adfm.202206753

[17]   A. Kumar, K. Gupta, S. Dixit, K. Mishra, S. Srivastava, A review on positive and negative impacts of nanotechnology in agriculture, Int. J. Environ. Sci. Technol. 16 (2019) 2175–2184. https://doi.org/10.1007/s13762-018-2119-7

[18]   H. Chung, Y. Son, T.K. Yoon, S. Kim, W. Kim, The effect of multi-walled carbon nanotubes on soil microbial activity, Ecotoxicol. Environ. Saf. 74 (2011) 569–575. https://doi.org/10.1016/j.ecoenv.2011.01.004

[19]   S. Dharni, Sanchita, S.M. Unni, S. Kurungot, A. Samad, A. Sharma, D.D. Patra, In vitro and in silico antifungal efficacy of nitrogen-doped carbon nanohorn (NCNH) against Rhizoctonia solani, J. Biomol. Struct. Dyn. 34 (2016) 152–162. https://doi.org/10.1080/07391102.2015.1018841

[20]   S. Kang, Y. Zhu, M. Chen, G. Zeng, Z. Li, C. Zhang, P. Xu, Can microbes feed on environmental carbon nanomaterials?, Nano Today. 25 (2019) 10–12. https://doi.org/10.1016/j.nantod.2019.01.001

[21]   F. Gamoń, A. Ziembińska-Buczyńska, D. Łukowiec, M. Tomaszewski, Ecotoxicity of selected carbon-based nanomaterials, Int. J. Environ. Sci. Technol. 20 (2023) 10153–10162. https://doi.org/10.1007/s13762-022-04692-w

[22]   M. V. Khodakovskaya, K. De Silva, D.A. Nedosekin, E. Dervishi, A.S. Biris, E. V. Shashkov, E.I. Galanzha, V.P. Zharov, Complex genetic, photothermal, and photoacoustic analysis of nanoparticle-plant interactions, Proc. Natl. Acad. Sci. U. S. A. 108 (2011) 1028–1033. https://doi.org/10.1073/pnas.1008856108

[23]   X. Yang, Z. Shen, B. Zhang, J. Yang, W.X. Hong, Z. Zhuang, J. Liu, Silica nanoparticles capture atmospheric lead: Implications in the treatment of environmental heavy metal pollution, Chemosphere. 90 (2013) 653–656. https://doi.org/10.1016/j.chemosphere.2012.09.033

[24]   X. Yang, Z. Shen, B. Zhang, J. Yang, W.X. Hong, Z. Zhuang, J. Liu, Silica nanoparticles capture atmospheric lead: Implications in the treatment of environmental heavy metal pollution, Chemosphere. 90 (2013) 653–656. https://doi.org/10.1016/j.chemosphere.2012.09.033

[25]   M.H. Sarfraz, M. Zubair, B. Aslam, A. Ashraf, M.H. Siddique, S. Hayat, J.N. Cruz, S. Muzammil, M. Khurshid, M.F. Sarfraz, A. Hashem, T.M. Dawoud, G.D. Avila-Quezada, E.F. Abd_Allah, Comparative analysis of phyto-fabricated chitosan, copper oxide, and chitosan-based CuO nanoparticles: antibacterial potential against Acinetobacter baumannii isolates and anticancer activity against HepG2 cell lines, Front. Microbiol. 14 (2023). https://doi.org/10.3389/fmicb.2023.1188743

[26]   H. Feizi, M. Kamali, L. Jafari, P. Rezvani Moghaddam, Phytotoxicity and stimulatory impacts of nanosized and bulk titanium dioxide on fennel (Foeniculum vulgare Mill), Chemosphere. 91 (2013) 506–511. https://doi.org/10.1016/j.chemosphere.2012.12.012

[27]   S. Jafari, D. Davoodi, P. Jonoubi, A. Majd, H. Alizadeh, Z.S. Shobbar, Anatomical assessment of maize (Zea mays L.) seedlings exposed to colloidal silver nanoparticles, Appl. Ecol. Environ. Res. 16 (2018) 2391–2401. https://doi.org/10.15666/aeer/1603_23912401

[28]   A. Mattiello, A. Filippi, F. Pošćić, R. Musetti, M.C. Salvatici, C. Giordano, M. Vischi, A. Bertolini, L. Marchiol, Evidence of phytotoxicity and genotoxicity in hordeum vulgare L. Exposed to CeO2 and TiO2 Nanoparticles, Front. Plant Sci. 6 (2015) 1–13. https://doi.org/10.3389/fpls.2015.01043

[29]   B. Singh Sekhon, Nanotechnology in agri-food production: An overview, Nanotechnol. Sci. Appl. 7 (2014) 31–53. https://doi.org/10.2147/NSA.S39406

[30]   W. Sun, F. Dou, C. Li, X. Ma, L.Q. Ma, Impacts of metallic nanoparticles and transformed products on soil health, Crit. Rev. Environ. Sci. Technol. 51 (2021) 973–1002. https://doi.org/10.1080/10643389.2020.1740546

[31]   M. Khan, M.S.A. Khan, K.K. Borah, Y. Goswami, K.R. Hakeem, I. Chakrabartty, The potential exposure and hazards of metal-based nanoparticles on plants and environment, with

special emphasis on ZnO NPs, TiO2 NPs, and AgNPs: A review, Environ. Adv. 6 (2021) 100128. https://doi.org/10.1016/j.envadv.2021.100128

[32]   P.C. Ray, H. Yu, P.P. Fu, Toxicity and environmental risks of nanomaterials: Challenges and future needs, J. Environ. Sci. Heal. - Part C Environ. Carcinog. Ecotoxicol. Rev. 27 (2009) 1–35. https://doi.org/10.1080/10590500802708267

[33]   J.N. Cruz, S. Muzammil, A. Ashraf, M.U. Ijaz, M.H. Siddique, R. Abbas, M. Sadia, Saba, S. Hayat, R.R. Lima, A review on mycogenic metallic nanoparticles and their potential role as antioxidant, antibiofilm and quorum quenching agents, Heliyon. 10 (2024). https://doi.org/10.1016/j.heliyon.2024.e29500

[34]   J.R. Conway, A.A. Keller, Gravity-driven transport of three engineered nanomaterials in unsaturated soils and their effects on soil pH and nutrient release, Water Res. 98 (2016) 250–260. https://doi.org/10.1016/j.watres.2016.04.021

Nanomaterials in Biological Systems   Materials Research Forum LLC
Materials Research Foundations 185 (2026) 164-182   https://doi.org/10.21741/9781644903858-8

Chapter 8

Ethical Considerations for Nanotechnology Applications in Biology

Naresh Kumar[1], Jaspreet Kaur[2], Brij Bala[1], Jibanananda Mishra[1]*

[1]School of Biosciences, RIMT University, Mandi Gobindgarh, Punjab147301, India

[2]Department of Medical Laboratory Science, RIMT University, Mandi Gobindgarh, Punjab147301, India

mjiban@gmail.com

Abstract

Nanotechnology has surfaced as a rapidly progressing field with numerous applications in biology, including molecular detection and manipulation, bio- and non-bio interface devices and materials, and drug delivery agents. However, the ethical considerations surrounding the applications of nanotechnology in biology are of paramount importance, as they encompass a range of issues. One of the key ethical considerations in nanotechnology is the potential impact on human health and the environment. The ethical implications of nanotechnology in biology have been a subject of debate for many years. The potential risks and benefits associated with different applications of nanotechnology and the need for responsible governance and oversight of nanotechnology in biology are major ethical considerations for researchers and policymakers. Furthermore, the social and cultural dimensions of nanotechnology ethics cannot be overlooked. Addressing questions related to malevolent uses, equitable access, control, and decision-making, as well as privacy and human integrity, are essential in ensuring responsible development and utilization of nanotechnology. By addressing these ethical considerations, we can promote the responsible and sustainable use of nanotechnology in biology. This chapter aims to provide an overview of the ethical considerations for nanotechnology applications in biology.

Keywords

Bioethics, Biotechnology, Nanotechnology, Nanomaterials

Contents

1.   Introduction

The scientific possibilities, as it implies, refers to anything that tends to offer room for advancement, as reflected by an article given by Richard Feynman (1959) titled "Plenty of room at the bottom"[1]. A path of development for the scientists in the following decades is somehow summarized in a very eloquent way, and it served as an informed assessment of the scientific potential of different technological advancement *viz*., nanotechnology.

Nanotechnology, as it implies, refers to the manipulation of matter's properties and behavior on an atomic and molecular scale in order to study systems at the nanoscale (one billionth of a meter). The field was subject to growing public awareness and controversy in the early 2000s, and in turn, the beginning of commercial applications of nanotechnology. It tends to be made possible with the best efforts, and the nanotechnology revolution is opening up a world of new possibilities. The field of nanotechnology encompasses several disciplines, including physics, chemistry, materials science, biology, and engineering primarily, since it is an interdisciplinary branch of science (Table 1).

Materials, structures, and devices at this scale need to be understood and manipulated in order to develop new applications and functionalities. Interestingly, unlike non-nanoscale particles with the same chemical composition, it has properties that are unique and different. In other words, particles with the same chemical composition that are nanoscale have different properties than those that are non-nanoscale [2,3]. Nanotechnology works at the nanoscale; it may be called nature's precious toy box due to the scope of creating new things with unlimited possibilities. It can be well known and exhibited through different nouns like nanomaterials, nano devices etc., and so forth. As an example, manipulation at nanoscale using technological advancements has led us to treat incurable diseases where it is impossible to see the nanostructures in human. On the contrary, the scientific community is concerned about them as well, since it is impracticable to predict the long- and short-term repercussions of all the use of this technological advancement [4].

Evitable, we are far from being in the realm of science fiction when it comes to nanotechnology today. Nanotechnology hauls knowledge and techniques from different disciplines to explore the unique properties and behaviors of nano-scale materials. A major advantage of nanotechnology is its ability to amalgamate scientific and technological information from diverse fields, which would lead to revolution in electronics, computing, energy, medicine, healthcare, environmental science, energy production and storage, environmental monitoring, and remediation, and others.

Table 1: Interdisciplinary branches in nanotechnology.

S. No.	Research Field	Examples
1	Microscopy	Scanning probe microscopy
2	Material Science	Nanoparticle research, Nanostructured materials
3	Surface Science	Polymers and composites ultra-thin coating
4	Organic Chemistry	Heterogeneous catalysis
5	Quantum Chemistry	Supramolecular chemistry, Hydrogen storage systems
6	Electrochemistry	Graphene materials, Metal nanoparticles
7	Electric Engineering	Molecular electronics
8	Solid State Physics	Molecular modelling lithography in the production of ITs (integrated circuits), Semiconductor research and quantum dots, Liquid crystals, Small LEDs (light emitting diodes), Solar cells
9	Mechanical and Chemical Engineering	Quantum computing MEMS (micro-electro-mechanical systems)
10	Biochemistry	Biochemical sensors
11	Molecular Biology	Targeted drug delivery, Molecular biotechnology, Genetic engineering
12	Physiology	Neurophysiology, Tissue engineering
13	Computing	Microcomputing

When any new technologies emerge, their core competencies flux, and in fact, they come with inherent merits and demerits that are challenging in the real world. Nanotechnology is no exception; perhaps it is beyond the scope of scientific possibilities and fantasy embracement but still within the reach of a scientific approach. It is also an imperative statement that concerns regarding nanotechnology are based on misconceptions. With such oblivious competencies, it somehow needs to have a framework for ethical concern for such an advanced science; however, safety and ethical aspects are either ignored or yet to be realized [5]. In any case, it is important for scientific committees at all levels to reflect on the bioethical implications of nanotechnology research. In a quest to understand how particles and molecules interact to create and configure certainty, researchers dive into the nanometric technological world with freedom of implausible research and an appetite to explore and learn from nature against the arguable and undiscovered science having risks of being existence [6]. This implies that nanotechnology research must proceed from a sense of innate need for knowledge to a greater understanding of its potential applications and risks.

The use and development of nanotechnology represent a range of benefits and risks for humans' social, economic, cultural, and even more complex lives. Nanotechnology, therefore, needs to be ethically considered both for its potential benefits and risks for society, the economy, culture, and even life itself. In order to accomplish this, a wide range of bioethical analysis proposals related to nanotechnology use have been developed. The use of nanotechnology has led to a variety of bioethical conflicts that are being analyzed in bioethical terms. Nanotechnology-related conflicts have thus been analyzed bio-ethically through various proposals [7]. Therefore, an exhaustive analysis of the literature is needed to explore the ethical and bioethical implications of various nanotechnology applications. An analysis of the literature should also include an assessment of the possible benefits and threats of the different nano-technological applications, as well as an analysis of their ethical and bioethical implications.

1.1 Nanoethics

Nanoethics is a field of study that focuses on the ethical and societal implications of nanotechnology. Nanotechnology involves manipulating matter at the nanoscale, which is the scale of atoms and molecules. It has the potential to revolutionize various industries, including medicine, electronics, and energy. However, along with its potential benefits, nanotechnology also raises ethical concerns and societal implications that need to be addressed [8].

One of the key ethical issues in nanotechnology is the potential risks associated with the use of nanomaterials. The unique properties of nanomaterials can pose health and environmental risks that are not well understood. Clinical trials involving nanomedicine present unique challenges related to risk minimization, management, and communication involving human subjects [9]. Risk perception is influenced by ethical concerns, and those who have ethical concerns with nanotechnology are more likely to perceive risk [10]. Another ethical concern in nanotechnology is the equitable distribution of its benefits. There is a concern that nanotechnology may exacerbate existing social inequalities if access to its benefits is limited to certain groups or countries. Fairness, justice, equity, and power are fundamental ethical concepts that unite the various social and ethical issues identified in nanotechnology [11]. Public perception and awareness of nanotechnology also play a significant role in nanoethics. Despite the amount of public investment in nanotechnology ventures, research shows that there is little public awareness about nanotechnology, and public knowledge is very limited [12]. The media plays a crucial role in shaping public perception of nanotechnology. An empirical analysis of newspaper coverage of nanotechnology suggests that newspapers often emphasize the positive aspects and scientific interpretations of nanotechnology, while downplaying potential risks and ethical concerns [13].

Public engagement and deliberation are important aspects of nanoethics. Engaging the public in debates about nanotechnologies can help ensure that their concerns and values are taken into account in decision-making processes. Nanotechnologies offer a site for intense future public engagement and deliberation [14], Including social and ethical implications in interdisciplinary nanotechnology courses can also raise awareness among students and prepare them for collaborative research [15].

The following social objectives should be aimed at and accomplished by stakeholders before any nano-technological product development [16].

- Launch of the nano-products in the society

- Pronouncement against the development of such technology

- The impact of nano-products on environment and human health

- Possible corollary to use of nano-products

1.2   Nanoparticles

As mentioned earlier, particles with a diameter of 1 to 100 nanometers (nm) are considered as nanoparticles or ultrafine particles. They may be either of natural or artificial origin. As defined by the International Organization for Standardization (ISO), nanoparticles are equivalent, or so-called nano-objects, whose longest and shortest axes are not significantly different in length and have all external dimensions at the nanoscale. Further, as nanoparticles often differ significantly in size (typically by a factor of three), it may be preferable to refer to them as nanofibers or nanoplates rather than nanoparticles. Nanoparticles exist in different shapes, sizes, and structures, *i.e.*, spherical, cylindrical, conical, tubular, hollow core, spiral, irregular, etc. It is usually preferable to speak of atom clusters when the size of NPs falls below 1 nm. In terms of the solid state of nanoparticles, they can either be crystalline single crystals, multi-crystal solids, or loose or agglomerated amorphous. Further, nanoparticles can exist uniformly or be composed of several layers that may have a variable composition, with the surface layer usually consisting of small molecules, metal ions, surfactants, or polymers, followed by the shell layer and the core layer at the central canal with different chemical identities in either of the layers [17-19].

1.2.1   Classification of nanoparticles

Nanoparticles are primarily categorized based on their composition into three classes: organic, inorganic, and carbon-based nanoparticles [18]. Further, the fourth class has been introduced due to its structural complexity, viz., composite nanoparticles [20].

1.2.1.1 Organic nanoparticles (ONPs)

There are many different organic compounds in this class, such as proteins, carbohydrates, lipids, polymers, and other organic compounds. A few prominent examples of current implications are cyclodextrin, dendrimers, liposomes, ferritin (protein-metal complexes), and micelles. Generally, organic nanoparticles are nontoxic, biodegradable, and can be sensitive to light and heat in some cases. Moreover, they are often formed by noncovalent interactions between molecules, which means they are more labile and can be cleared from the body by the body's channels [21, 18]. Currently, organic nanoparticles are being used in targeted drug delivery and cancer [22, 18].

1.2.1.2 Inorganic nanoparticles

There is a broad variety of inorganic nanoparticles, to name a few, such as metal, ceramic, and semiconductor nanoparticles, in this class. Metal NPs are made purely of metal precursors. They can be monometallic, bimetallic, or polymetallic. Bimetallic nanoparticles are made of alloys or formed in layers (core-shell). These nanoparticles have unique optical and electrical properties because of their localized surface plasmon resonance properties. Additionally, metal nanoparticles have distinct thermal, magnetic, and biological properties. These properties make them great materials for creating nanodevices for a variety of applications in chemistry, biology, biomedicine, and pharmaceuticals. Metal nanoparticles are important for creating cutting-edge materials [18–19, 23-24]. Semiconductor nanoparticles are made of materials that show properties equivalent to those of metals and non-metals. So, they can be used in photocatalysis, optics, and electronic devices. Ceramic nanoparticles are composed of metals (titanium and calcium) in juxtaposition with carbonates, carbides, phosphates, and oxides. Amorphous polycrystalline, dense porous, or hollow forms can be obtained by heating and cooling successively. They are primarily used in biological applications due to their high stability and high load capacity, but they are also used in other applications such as catalysis, degradation of dyes, photonics, and optoelectronics [25-26].

1.2.1.3 Carbon-based nanoparticles

Carbon-based nanoparticles are exclusively composed of carbon atoms. These cylindrical carbon structures have unique mechanical, electrical, and thermal properties. The well-known examples of carbon-based nanoparticles are fullerenes, carbon-black nanoparticles, and carbon quantum dots.

1.2.1.4 Other types of nanoparticles

Nanoparticles come in various types based on their composition, size, and shape. Here are some common types of nanoparticles congenitally illustrated:

Metal nanoparticles

Gold nanoparticles: Gold nanoparticles are widely used in applications such as diagnostics, drug delivery, and cancer therapy due to their unique optical properties and biocompatibility.

Silver nanoparticles: Silver nanoparticles are known for their antimicrobial properties and find use in wound dressings, textiles, and water purification.

Metal oxide nanoparticles

Iron oxide nanoparticles: These are used in magnetic resonance imaging (MRI) contrast agents and as drug carriers in targeted therapies.

Titanium dioxide (TiO_2) nanoparticles: TiO_2 nanoparticles are used in sunscreens, cosmetics, and as photocatalysts for environmental applications.

Zinc oxide (ZnO) nanoparticles: ZnO nanoparticles have UV-blocking properties and are used in sunscreens, cosmetics, and as antibacterial agents.

Silicon oxide (SiO_2) nanoparticles: SiO_2 nanoparticles are used in drug delivery, as a food additive, and in the development of nanocomposites.

Carbon-based nanoparticle

They are used in electronics, composites, and nanomedicine.

Fullerenes: Spherical carbon molecules, like Buckminsterfullerene (C_{60}), are used in research for drug delivery and as antioxidants

Organic nanoparticles

*Polymeric nanoparticles (nanospheres and nanocapsules):*These nanoparticles are made from polymers like PLGA (polylactic-co-glycolic acid) and are used in drug delivery systems

Dendrimers: These highly branched polymer nanoparticles have applications in drug delivery and diagnostics.

Lipid-based nanoparticles (Liposomes): These are spherical vesicles composed of lipid bilayers. Liposomes are used to encapsulate and deliver drugs and genes.

Solid-lipid nanoparticles (SLNs): SLNs are used for drug delivery and have advantages like controlled release and enhanced bioavailability

*Inorganic nanoparticle (Quantum dots):*These semiconductor nanoparticles have unique optical properties and are used in imaging, displays, and as biological markers.

Metal-organic frameworks (MOFs): MOFs are porous materials made of metal ions connected by organic linkers and are used in gas storage, catalysis, and drug delivery.

Magnetic nanoparticles: These nanoparticles contain magnetic materials and are used in applications like magnetic resonance imaging (MRI), targeted drug delivery, and hyperthermia cancer treatment

Biological nanoparticles (virus-like particles-VLPs): These self-assembling biological nanoparticles are used in vaccine development and drug delivery

Exosomes: Natural nanoparticles secreted by cells; they have potential applications in drug delivery and diagnostics.

Hybrid nanoparticles: These nanoparticles combine different materials or types, providing unique properties and versatile applications. For example, silica-coated magnetic nanoparticles combine the properties of both materials for various applications.

These are just a few examples of the many types of nanoparticles that exist. The choice of nanoparticle type depends on the specific application and the desired properties, such as size, shape, and material characteristics. Nanoparticles continue to be a vibrant area of research and innovation with the potential to revolutionize various industries.

1.3   Nano-ecotoxicology

The science of studying the effects that refer to "pesticides and other contaminants that get into the natural environment" and affect plants and animals, humans, and even microorganisms, including bacteria, fungi, algae, etc., is called ecotoxicology. Ecotoxicology is a complex juxtaposition between ecology, toxicology, physiology, analytical chemistry, molecular biology, and mathematics. Nanomaterials are prepared or manufactured in order to compensate for the current implications in biomedical or other related or non-related fields where nanomaterials have replaced the materials at work. Many commercially available consumer products contain manufactured nanomaterials, including cosmetics, textiles, and paints. As these materials are produced in greater quantities, their environmental impact becomes more evident. To list examples, the toxicological impact of some nanomaterials, such as ZnO, TiO_2, and $BaTiO_3$ nanoparticles, on aquatic microorganisms can be incidentally discussed.

In the same way, nano-ecotoxicology, as being named after the effect of nanomaterials on ecosystems such as ponds, rivers, deserts, grasslands, and forests, looks at the impacts of nano-contaminants, including organic, inorganic, or carbon-based nanoparticles, on individuals, populations, natural communities, and ecosystems. Ecotoxicologists also study what happens to the pesticides themselves, where they go in the environment, how long they last, and how they finally break down. In recent years, the number of toxicological studies has increased to see the effects of hazards posed by manufactured nanoparticles. Scientists are more concerned about the mode of action and mechanisms of toxic action of nanoparticles. An argument is centered on the perception of which inherent and physical characteristics of nanoparticles considerably contribute to adverse effects. Predictive models are being developed to investigate the relationships between the intrinsic properties of nanoparticles and their biological impacts to reduce the need for costly and lingering experimental testing [27-29].

2. Intervention support for ethical issues

At this moment, everyone is focused on discovering the positive aspects of nanoparticles, but no one seems to identify the detrimental impact of these small particles on the health and environment. Due to their nanoscale properties, these nanoparticles may be able to migrate across biomembranes and display catalytic effects in addition to their chemical, thermodynamic, optical, mechanical, electromagnetic, and biological properties.

2.1 Health issues

It was believed that inhaling nanoparticles may cause particles to accumulate inside the lungs and act as a mediator of cellular oxidative stress and inflammatory reactions in the body. This indirectly affected the cardiovascular system's ability to function effectively. It was proven that extrapulmonary effects were stimulated by carbon nanotube inhalation. This was explained by direct consequences, namely carbon nanotube translocation, as well as additional impacts from particle-cell interactions in the lungs. A similar study has shown that multi-wall carbon nanotubes being exposed lead to depositing in the respiratory system before they circulate in singlet form to the parietal pleura, muscles of respiration, kidneys, liver, heart, and brain [30]. In addition to these detrimental effects, nanotechnology exposure has been linked to minor allergic responses, malignancies of the lungs, and cirrhosis. The size, shape, and dosage of nanomedicine could decide whether nanoparticle exposure is negative or not. One example is TiO_2 nanoparticles, which have been linked to lung fibrosis and inflammatory conditions such as asthma and chronic obstructive pulmonary disorders [31].

Nanoparticles can pass the blood-brain barrier because of their small size. These particles could increase their concentrations in the brain and resist phagocytosis. Polystyrene nanoparticles (PS-NPs), for example, have been shown in zebrafish research to trigger inflammation in the intestinal tract, developmental constraints, and growth reduction. This has been linked to a disturbance in the brain-gut-microbiota system. Nanoparticle exposure, including silicon dioxide, iron oxide, silver, lead oxide, and ZnO, may adversely affect the brain. Every particle has different mechanism, but they all enter the body through the nasal canals, where they trigger oxidative stress, disrupt the cell cycle, and ultimately cause damage to the nervous system [31].Exposure to such particles increase levels of ROS (Reactive Oxygen Species), which might have negative consequences such as neurological problems and oxidative stress. Metal oxide nanoparticles have the ability to pass the blood-brain barrier and concentrate in the olfactory bulb, cerebellum, cortex, or other areas of the brain [32].

There are some historical incidents related to the exposure of nanomaterial as shown in the Table 2:

Nanomaterials in Biological Systems   Materials Research Forum LLC
Materials Research Foundations 185 (2026) 164-182   https://doi.org/10.21741/9781644903858-8

Table 2: Nanoparticles with their harmful effects

S.No.	Nanoparticles/ Nanomaterials	Harmful effects
1.	Copper oxide nanoparticles (CuONPs)	The apoptosis rate increases with the fragmentation of DNA [31]. By blocking sodium current, it may adversely affect memory and learning [31]. It was experimentally proven by Mahmoud et al. (2016) that cells could be dead in the SHSY5Y cell line by exposure to cupric oxide [33].
2.	Silver nanoparticles	Overexposure to particles may result in enhanced neurodegenerative and astrocyte responses. This results from blood-brain barrier disruption with cerebral microvasculature [31, 34]. Increases allergic reactions or inflammatory responses [31, 34]. Interrupt the growth of neurons and the cell multiplication process [31]. It is proven to be potentially damaging to the genome by the formation of micronuclei due to reactive oxygen species formation [32].
3.	Silica nanoparticles	Density of cells decreased [31]. Dendrite processes are also slower [31]. Alzheimer's disease could occur due to an increase in the $A\beta1-42$ peptide [31].
4.	Magnetic carriers(used as a nanocarrier to target drug delivery across the blood-brain barrier)	This carrier goes through various processes for stability and can work effectively with iron particles that are hazardous to the brain and cells by causing apoptosis [31].
5	Silicon-coated nano-capsule	The primary way to cross the blood-brain barrier is to interrupt processes of cell membrane translocation. This causes the blood-brain membrane to remain permeable for an extended period of time, which allows a number of impermeable substances to pass through[31].
6	Zinc oxide nanoparticles	In experiments on rats, it was found that it has an impact on memory and learning capacity. It can cause DNA breakage and the apoptotic response in cells by inducing ROS in macrophages [35].

Beyond the nervous system, these particles may also affect the reproductive system. As for the blood-brain barrier, these particles can pass across the blood-testis barrier, epithelial barrier, or placental barrier and get deposited in the reproductive organs (ovary, uterus, epididymis, and testis). The damage of Leydig, Sertoli, and germ cells by nanoparticle deposition could result in impaired reproductive system function. As a consequence, sperm quality, morphology, quantity, and motility may be reduced, or the number of mature oocytes may decrease, and primary and secondary follicular development may be disrupted. Apart from this, hormones can also be affected. For instance, a study found that gold alloy caused epididymis inflammation, which further reduced the sperm motility [36]. Cerium oxide nanoparticles and silver nanoparticles modified seminiferous tubules too. Furthermore, cerium oxide nanoparticles have been reported to significantly reduce levels of testosterone, follicle-stimulating hormone (FSH), luteinizing hormone (LH), and prolactin [37], Examples of nanoparticles involved in female reproductive system function impairment include TiO_2 particles, which decrease the number of mature oocytes by modifying follicle morphological characteristics. The particles accumulate in the ovary, triggering apoptosis, which has an adverse impact on fertility. Likewise, hormone levels become unsteady as progesterone (P4), LH, and testosterone (T) levels may fall while estradiol (E2) and FSH levels increase [38, 31].

Besides this, entire endocrine system is also reported to be affected by nanoparticles. Firstly, it is observed that nanoparticles could mimic hormones. The mechanism behind this is that nanoparticles can attach to hormone receptors and then trigger or block the signal cascade. Silver-zinc oxide is taken as an example of this process. These particles were observed to imbalance the endocrine system, especially in the case of pregnant females and kids, because they attach to the oestrogen hormone receptors to start oestrogenic activity. Interruption in the signalling cascade is another way to imbalance the endocrine system by inducing oxidative stress or ROS. For instance, carbon nanoparticles preventing aromatase enzymes are essential for oestrogen biosynthesis, which leads to lower oestrogen levels and irregularities in endocrine functioning [39].

Additionally, nanoparticles have significant effects on the immune system. Nanoparticles have been shown to trigger innate immunological reactions when they are identified by the immune system. The evidence of this includes reduced hemophagocytosis and minimized or inactive immunological reactions in *Crassostrea gigas* (oyster) [40]. According to another investigation, cadmium nanoparticle exposure in mice decreased the viability of monocyte cells. Aside from PS-NPs, ZnO-NPs, Ag-NPs, and Au-NPs, other nanoparticles are also responsible for triggering innate immunity; however, they don't result in inflammatory reactions [41]. Further, certain kinds of nanoparticles may eventually cause inflammatory or immunological reactions [42].

Moreover, nanoparticles cause cancerous or fatally malignant reactions in humans. It is well known that cancer has the greatest mortality rate worldwide and is one of the most fatal diseases. Therefore, researchers and healthcare providers are always exploring novel strategies to entirely eradicate cancer from the world. Carcinogenesis is a condition that involves initiating, ascending, and progressing and causes normal cells to divide and transform into cancer cells. In the initial phase, where there are only slight genetic modifications, benign tumors are formed. Malignancy, on the other hand, is referred to as slightly greater genetic alteration. Nanoparticles may cause DNA damage by apoptosis, a gene mutation, elevated chemokine or cytokine expression, cell cycle inhibition, immune suppression, inflammation, or reduced viability of important cell types that play a vital role in the immune system. Metal nanoparticles, such as ZnO, hexavalent chromium, and nickel nanoparticles, are thought to trigger a carcinogenic response [31].

2.2   Environmental issues

Nanoparticle-based products have been manufactured by various industries since they think that these products possess trustworthy qualities that make their jobs easier. The negative consequences of the product waste or recycled materials would not, however, be given much weight due to a lack of awareness among the public. Nanoparticles are present in many essential goods, including paints, coatings, cosmetics, and catalytic additives. The nanoparticles are released in three distinctive ways, beginning with the raw material manufacturing procedure. Second, during the entire process. Finally, it is released into the environment post-usage or as waste material. Besides, the effluent of wastewater treatment plants or landfill leaches may pollute the earth in an indirect way [36]. Nanoparticles are found in nature as well as being created artificially.As compared to naturally occurring particles, man-made particles are more hazardous to the environment and human health [43]. Man-made nanoparticles have adverse impact on aquatic mammals or marine species. For example, cobalt or iron nanoparticles are used in remediation processes. An increase in these particles' concentration has been found to result in a drop in the algae's chlorophyll a and b. As a result, the surface and structure of the algae deteriorated, leading to the release of nano-oxide and chlorophyll, which disrupted the entire food chain and messed with the food chains of aquatic animals. Also vulnerable are the tissues of aquatic organisms [44].

Hydroxyl radicals are reactive oxygen radicals that break down organic matter and pollutants. Their depletion in the atmosphere causes an increase in greenhouse gases, the depletion of the ozone layer, and environmental damage. In addition, they increase exposure to UV radiation, which causes skin cancer. Additionally, the interaction between molecular hydrogen derived from hydrogen fuel cells and various sources and nanoparticles present in the troposphere results in the production of water vapour and subsequent cooling of the stratosphere. These phenomena result in the depletion of ozone, the formation of noctilucent clouds, and alterations in troposphere chemistry. As a consequence, they have significant effects on the interactions between the atmosphere and biosphere and may potentially exert an influence on climate variations [45].

2.3 Societal or individual effects of nanotechnology

Nanotechnology has improved some amazing benefits, but societal issues cannot be neglected. First and foremost, it would cause inequality or discrimination among people. It will especially have effects on developing countries. These countries spend the majority of their budgets to accomplish the needs of their citizens or to provide necessities to their residents. At this time, when wealthy nations are using nanotechnology to treat patients, developing nations are looking at how to fund this type of medical technology.

Researchers, workers, employees, or laborer who work on nanotechnology to manufacture various products for firms, knowingly or unknowingly endanger their lives. The waste materials may cause several kinds of fatal diseases too [46]. Moreover, a layman does not know about the negative implications of nanotechnology because it is not understandable by the general public. Makeup products may be taken as an example; the consumer uses these products without knowing about the precise dosage. It would adversely affect the lifestyle of an individual. This novel technology helps with many tasks performed by human beings. Besides this, the government also intrudes on the privacy of society or people by using the latest nanotechnology in the form of installing cameras, listening devices, face recognition devices, and tracking devices. These are beneficial to organizations or governments but have been directly imposed on the public, and people have no rights to avoid such surveillance [47-48]. Nanotechnology is also used by the government in the military to make advanced weapons, but it backfires on civilians if used by terrorists or politicians for their own purposes.

Apart from this, some critics of nanotechnology stated that life expectancy could be increased through the use of nanotechnology, and the older generation would live a longer period of time. However, this could lead to a decrease in the useful resources on earth or an increase in the living costs for new generations [33]. Finally, nanotechnology could help individuals change their identification by helping them make modifications in their appearance, body structure, performance, or even personality [49].

3.   Precautions and Regulation in Nanotechnology

The emergence of new products of nanotechnology has greatly ameliorated the standard of our day-to-day lives but has also raised concerns about their safety. There have been considerable national and international activities to examine its regulatory issues. The role of regulating nanotechnology is to diminish its risk to employees, end users, and the environment. It also strengthens public confidence in nanotechnology by providing commitment [50-51]. In the European Union, there are three bodies of cooperation affiliated with regulating nanotechnology, which include one legislative body, the European Chemical Agency (ECHA), and two non-legislative bodies, the European Food Safety Authority (EFSA) and the Scientific Committee on Emerging and Newly Identified Health Risks (SCENIHR). ECHA executes new material regulations by collecting and evaluating information on chemicals [52]. EFSA researches food, feed safety, animal, and plant protection and analyses the risk assessment. It contributes to delivering advice to policymakers [53]. SCENIHR utilizes health and environmental risk assessment to recognize and disseminate information on complicated or interdisciplinary topics [54]. In the United States, various agencies are involved in the regulation of nanotechnology products. These include the National Institute for Occupational Safety and Health (NIOSH), the Occupational Safety and Health Administration (OSHA), the Consumer Product Safety Commission (SPSC), the Food and Drug Administration (FDA), and the US Environmental Protection Agency (EPA). The FDA and EPA are the major stakeholders in regulating most of the products [51, 55-56]. The safety of food and feed additives that are positioned on the market under the Federal Food, Drug, and Cosmetic Act (FFDCA) is regulated by the FDA. Substances used as food and color additives are always subject to pre-market authorization by the FDA, while food ingredients that are generally recognized as safe (GRAS) don't require pre-market approval [57-58]. In 2009, the Nanomaterials Research Strategy was launched (2009–2014) by the EPA to gather data to establish future regulations. Its prime focus is to look at the distinctive properties of nanomaterials to develop forecasting algorithms and find more risky and safe nanomaterials [57]. In the US Environmental Regulations for Nanomaterials, there are provisions for different acts to identify, dispose of, control, distribute, and sell nanomaterial-related items. For instance, the Toxic Substances Control Act (TSCA) identifies and controls toxic substances; the Federal Insecticide, Fungicide, and Rodenticide Act (FIFRA) controls the distributing, selling, and use of nanomaterial-based pesticides. The Clean Air Act (CAA) measures nanoaerosols diffused from mobile devices into the air. The Clean Water Act (CWA) regulates the disposal of contaminating materials into waters, and the Resource Conservation and Recovery Act (RCRA) regulates the waste management of nanomaterials. In India, guidelines for regulating the usage of nanotechnology in the agriculture sector have been drafted by the Department of Biotechnology (DBT) under the Union Ministry of Science and Technology in 2020. This would accelerate technology commercialization by maintaining high standards of quality and security in products like nanofertilizers and nanopesticides [18]. These guidelines are applied to agri-input products in nanoform (inorganic/organic/composite, finished food formulations, finished feed formulations, nanocarriers, sensors, nanocomposites for food packaging, dairy products, etc. The guidelines aligned with the Food Safety and Standards Authority of India (FSSAI), the Bureau of Indian Standards (BIS), and other food and feed agencies. The guidelines proposed for nanofertilizers should be conducted under FCO (Fertilizer Control Order) 1985. FCO is issued under the Essential Commodities Act of 1955. Nanopesticide testing should be conducted as per the regulatory aspect of the Insecticides Act, 1968. Manufacturers and researchers who contribute to the expansion and advancement of nano-based commodities might find these guidelines beneficial. Certainly, this would benefit the agribusiness and food industries by providing a clear road map for financing organizations and by easing research interest while preserving practical safety procedures.

Risk assessment is crucial and should be conducted on a case-by-case basis by making use of appropriate data in the nanomaterial [51]. To minimize the risks of technology-related damage while attempting to diminish any disruptions or refusals of the advantages of the same technology, it is imperative to apply a modest amount of precaution. The implementation of the validated framework is currently the main obstacle, as are instruments for the identification, evaluation, and characterization of nanomaterials and tools to measure the exposure of nanomaterials. When there are potential threats to the health of people, animals, plants, or the environment, the precautionary principle enables prompt action. In Europe, there are different precautionary principles. In steward systematization, regulation should not be applied without scientific uncertainty about activities, and precautionary measures should comprise a safety margin, restricting activity to the point where no negative effects have been found. The use of the best available technology and activities with an unclear potential to cause severe harm should be prohibited. PrecauPri (Precautionary Principle), an EU-funded project through STRATA Programme, formulated a set of standards for comprehending and implementing the precautionary principle in practice within the European Union. A general model was established within the framework of the PrecauPri project to address the four main problems with modern risks: seriousness, uncertainty, complexity, and ambiguity [59]. According to the Rio Declaration, "States shall usually adopt a precautionary approach, following their capacities". Lack of complete scientific assurance shall not be used as an excuse for delaying cost-effective steps to avoid environmental deterioration where there are dangers of major or irreparable consequences. According to the North-East Atlantic, there would be good reasons to be apprehensive that energy or substance introduction into the environment, whether directly or indirectly, could impair human health or living things [60]. In 2000, the European Union issued a communication on the precautionary principle to indicate guidelines and understanding for applying the principle. The measures should be implemented when early scientific analysis shows that there are good reasons to be concerned about the possibly hazardous impacts on the environment. The aim of the measures is sustainable development, and they emphasize how choices made now will affect the welfare of future generations. The precautionary principle should be used in a transparent, non-discriminatory, and balancing manner. The three components of risk analysis—the evaluation of risk, the selection of a risk management plan, and the communication of risk—would be required for a systematic decision-making process with thorough scientific and other objective information. When the evidence is ambiguous and uncertain and there are signs that potential consequences on the environment, human, animal, or plant health may be dangerous and inconsistent with the level of protection chosen, the precautionary principle would be applied. Some factors trigger recourse to the precautionary principle that the usage of this principle is applicable only in the occurrence of a possible danger, in which it is necessary to identify any potential negative effects of a phenomenon with relevant scientific data evaluation. When determining if environmental protection measures are required, among other things, a scientific evaluation of the potential negative impacts should be conducted based on the evidence currently available. The risk assessment should be taken into account and should indicate the likelihood of happening, the impact of a hazard, the level of potential damage, persistency, reversibility, and any potential delayed effects. When adopting measures based on the evaluators' expert advice, risk managers should be fully aware of these uncertainty elements. The evaluation should determine whether the "desired level of protection" contains a summary of the hypotheses used to make up for the lack of scientific or statistical data and include an assessment of scientific uncertainty. Before releasing goods or substances onto the market, the precautionary principle is applied in several regulatory frameworks [59].

Conclusion

Ethical considerations are of utmost importance when it comes to the application of nanotechnology in biology. Nanotechnology has the potential to revolutionize various fields, including medicine, pharmacology, and tissue engineering. However, it also raises ethical concerns that need to be addressed to ensure responsible and safe use of this technology. One of the key roles of bioethics is to identify and explore the ethical problems associated with emerging technologies, such as nanotechnology. This is particularly relevant in the field of biology, where nanotechnology can have significant implications. For example, in the context of pharmacogenetics, ethical issues related to safety, toxicity hazards, and physiological effects need to be considered. The use of nanoparticles in drug delivery systems also raises concerns about safety and bioethical issues.

Consumer perceptions of nanotechnology applications in biology are influenced by a range of socio-psychological and affective factors, including ethical concerns. People may have ambiguous and pessimistic views about nanotechnology applications in the food domain, highlighting the need to address ethical considerations. Besides, societal acceptance of nanotechnology is influenced by factors such as risk assessment, business interests, and religiosity.

When it comes to tissue engineering, ethical considerations are linked to various aspects, including the use of human embryonic stem cells, privacy, and respecting privacy. The ethical dimensions of nanotechnology development are also a focus of governmental support in different regions, such as Japan, Europe, and the USA.

In the field of nanomedicine, ethical issues related to privacy, security, and human enhancement have been identified. The responsible governance of nanotechnology is crucial for its societal development, and ethical and legal aspects need to be considered. Furthermore, international collaboration regarding ethics for nanotechnology is required. Environmental ethics can guide the evaluation of ethical issues in nanotechnology, particularly in terms of its impact on the environment. Incorporating ethical concepts into university science education can ensure that social and ethical implications of nanotechnology research are taken into consideration. The role of ethics in the social implications of nanotechnology should be given due consideration.

Thus, ethical considerations play a vital role in the application of nanotechnology in biology. Addressing ethical concerns related to safety, privacy, human enhancement, and environmental impact is crucial for the responsible and sustainable development of nanotechnology. Bioethics and societal acceptance of nanotechnology are important areas of research that can contribute to the ethical discourse surrounding this emerging technology. It is crucial for researchers, policymakers, and society as a whole to engage in ongoing dialogue and reflection to ensure that nanotechnology in biology is developed and used in an ethically responsible manner.

References

[1]F. Nowak, Richard Feynman: There's Plenty of Room at the Bottom 1960 [1959], Nanotechnologie Als Kollekt. (2021) 459–470. https://doi.org/10.1515/9783839438039-022

[2]M. Auffan, J. Rose, J.Y. Bottero, G. V. Lowry, J.P. Jolivet, M.R. Wiesner, Towards a definition of inorganic nanoparticles from an environmental, health and safety perspective, Nat. Nanotechnol. 4 (2009) 634–641. https://doi.org/10.1038/nnano.2009.242

[3]C. Chellaram, G. Murugaboopathi, A.A. John, R. Sivakumar, S. Ganesan, S. Krithika, G. Priya, Significance of Nanotechnology in Food Industry, APCBEE Procedia. 8 (2014) 109–113. https://doi.org/10.1016/j.apcbee.2014.03.010

[4]National plan for ocean surveys / Interagency Committee on Oceanography of the Federal Council for Science and Technology., The Committee, 2011. https://doi.org/10.5962/bhl.title.39057

[5]I. van de Poel, How should we do nanoethics? A network approach for discerning ethical issues in nanotechnology, Nanoethics. 2 (2008) 25–38. https://doi.org/10.1007/s11569-008-0026-y

[6]E.N. Rogotneva, I. Melik-Haikazyan, M. Goncharenko, Bioethics: Negotiation of Fundamental Differences in Russian and US Curricula, Procedia - Soc. Behav. Sci. 215 (2015) 26–31. https://doi.org/10.1016/j.sbspro.2015.11.569

[7]M. Grant, Derechos humanos y trabajo social: alcances y límites del Código de Ética del Colegio Profesional de Servicio Social de Neuquén, Universidad Nacional de La Plata, n.d. https://doi.org/10.35537/10915/45850

[8]R.N. AlKahtani, The implications and applications of nanotechnology in dentistry: A review, Saudi Dent. J. 30 (2018) 107–116. https://doi.org/10.1016/j.sdentj.2018.01.002

[9]D.B. Resnik, S.S. Tinkle, Ethical issues in clinical trials involving nanomedicine, Contemp. Clin. Trials. 28 (2007) 433–441. https://doi.org/10.1016/j.cct.2006.11.001

[10]   S. Larsson, M. Jansson, Å. Boholm, Expert stakeholders' perception of nanotechnology: risk, benefit, knowledge, and regulation, J. Nanoparticle Res. 21 (2019). https://doi.org/10.1007/s11051-019-4498-1

[11]   B. V. Lewenstein, What counts as a "social and ethical issue" in nanotechnology?, Hyle. 11 (2005) 5–18. https://doi.org/10.4324/9781003075028-4

[12]   R. Kyle, S. Dodds, Avoiding empty rhetoric: Engaging publics in debates about nanotechnologies, Sci. Eng. Ethics. 15 (2009) 81–96. https://doi.org/10.1007/s11948-008-9089-y

[13]   B. Groboljsek, F. Mali, Daily newspapers' views on nanotechnology in Slovenia, Sci. Commun. 34 (2012) 30–56. https://doi.org/10.1177/1075547011427974

[14]   P. Macnaghten, Researching technoscientific concerns in the making: Narrative structures, public responses, and emerging nanotechnologies, Environ. Plan. A. 42 (2010) 23–37. https://doi.org/10.1068/a41349

[15]   E. Hoover, P. Brown, M. Averick, A. Kane, R. Hurt, Teaching Small and Thinking Large: Effects of Including Social and Ethical Implications in an Interdisciplinary Nanotechnology Course, J. Nano Educ. 1 (2011) 86–95. https://doi.org/10.1166/jne.2009.013

[16]   J.A. Khan, R.K. Kainthan, M. Ganguli, J.N. Kizhakkedathu, Y. Singh, S. Maiti, Water soluble nanoparticles from PEG-based cationic hyperbranched polymer and RNA that protect RNA from enzymatic degradation, Biomacromolecules. 7 (2006) 1386–1388. https://doi.org/10.1021/bm050999o

[17]   S. Machado, J.G. Pacheco, H.P.A. Nouws, J.T. Albergaria, C. Delerue-Matos, Characterization of green zero-valent iron nanoparticles produced with tree leaf extracts, Sci. Total Environ. 533 (2015) 76–81. https://doi.org/10.1016/j.scitotenv.2015.06.091

[18]   I.N. de F. Ramos, M.F. da Silva, J.M.S. Lopes, J.N. Cruz, F.S. Alves, J. de A.R. do Rego, M.L. da Costa, P.P. de Assumpção, D. do S. Barros Brasil, A.S. Khayat, Extraction, Characterization, and Evaluation of the Cytotoxic Activity of Piperine in Its Isolated form and in Combination with Chemotherapeutics against Gastric Cancer, Molecules. 28 (2023). https://doi.org/10.3390/molecules28145587

[19]   I. Khan, K. Saeed, I. Khan, Nanoparticles: Properties, applications and toxicities, Arab. J. Chem. 12 (2019) 908–931. https://doi.org/10.1016/j.arabjc.2017.05.011

[20]   G. Luo, L. Du, Y. Wang, K. Wang, Composite Nanoparticles, Encycl. Microfluid. Nanofluidics. (2014) 1–9. https://doi.org/10.1007/978-3-642-27758-0_243-3

[21]   K.K. Ng, G. Zheng, Molecular Interactions in Organic Nanoparticles for Phototheranostic Applications, Chem. Rev. 115 (2015) 11012–11042. https://doi.org/10.1021/acs.chemrev.5b00140

[22]   M. Gujrati, A. Malamas, T. Shin, E. Jin, Y. Sun, Z.R. Lu, Multifunctional cationic lipid-based nanoparticles facilitate endosomal escape and reduction-triggered cytosolic siRNA release, Mol. Pharm. 11 (2014) 2734–2744. https://doi.org/10.1021/mp400787s

[23]   V. Mody, R. Siwale, A. Singh, H. Mody, Introduction to metallic nanoparticles, J. Pharm. Bioallied Sci. 2 (2010) 282. https://doi.org/10.4103/0975-7406.72127

[24]   E.C. Dreaden, A.M. Alkilany, X. Huang, C.J. Murphy, M.A. El-Sayed, The golden age: Gold nanoparticles for biomedicine, Chem. Soc. Rev. 41 (2012) 2740–2779. https://doi.org/10.1039/c1cs15237h

[25]   R. D'Amato, M. Falconieri, S. Gagliardi, E. Popovici, E. Serra, G. Terranova, E. Borsella, Synthesis of ceramic nanoparticles by laser pyrolysis: From research to applications, J. Anal. Appl. Pyrolysis. 104 (2013) 461–469. https://doi.org/10.1016/j.jaap.2013.05.026

[26]   S. Thomas, B.S.P. Harshita, P. Mishra, S. Talegaonkar, Ceramic Nanoparticles: Fabrication Methods and Applications in Drug Delivery, Curr. Pharm. Des. 21 (2015) 6165–6188. https://doi.org/10.2174/1381612821666151027153246

[27]   V. Pachapur, S.K. Brar, M. Verma, R.Y. Surampalli, Nano-ecotoxicology of natural and engineered nanomaterials for animals and humans, Nanomater. Environ. (2015) 421–438. https://doi.org/10.1061/9780784414088.ch16

[28]   F.S. Alves, J.N. Cruz, I.N. de Farias Ramos, D.L. do Nascimento Brandão, R.N. Queiroz, G.V. da Silva, G.V. da Silva, M.F. Dolabela, M.L. da Costa, A.S. Khayat, J. de Arimatéia Rodrigues do Rego, D. do Socorro Barros Brasil, Evaluation of Antimicrobial Activity and Cytotoxicity Effects of Extracts of Piper nigrum L. and Piperine, Separations. 10 (2023). https://doi.org/10.3390/separations10010021

[29]   NHS managers need support to use tools to ensure safe nurse staffing levels, National Institute for Health and Care Research, 2019. https://doi.org/10.3310/signal-000724

[30]   Z. Wang, M.G. Vijver, W.J.G.M. Peijnenburg, Multiscale Coupling Strategy for Nano Ecotoxicology Prediction, Environ. Sci. Technol. 52 (2018) 7598–7600. https://doi.org/10.1021/acs.est.8b02895

[31]   J.N. Cruz, S. Muzammil, A. Ashraf, M.U. Ijaz, M.H. Siddique, R. Abbas, M. Sadia, Saba, S. Hayat, R.R. Lima, A review on mycogenic metallic nanoparticles and their potential role as antioxidant, antibiofilm and quorum quenching agents, Heliyon. 10 (2024). https://doi.org/10.1016/j.heliyon.2024.e29500

[32]   L. Xuan, Z. Ju, M. Skonieczna, P.K. Zhou, R. Huang, Nanoparticles-induced potential toxicity on human health: Applications, toxicity mechanisms, and evaluation models, MedComm. 4 (2023). https://doi.org/10.1002/mco2.327

[33]   C. Vinod, S. Jena, Nano-Neurotheranostics: Impact of Nanoparticles on Neural Dysfunctions and Strategies to Reduce Toxicity for Improved Efficacy, Front. Pharmacol. 12 (2021). https://doi.org/10.3389/fphar.2021.612692

[34]   M. Abudayyak, E. Guzel, G. Ozhan, Copper (II) oxide nanoparticles induce high toxicity in human neuronal cell, Toxicol. Lett. 258 (2016) S262–S263. https://doi.org/10.1016/j.toxlet.2016.06.1925

[35]   K. Dziendzikowska, J. Wilczak, W. Grodzicki, J. Gromadzka-Ostrowska, M. Węsierska, M. Kruszewski, Coating-Dependent Neurotoxicity of Silver Nanoparticles—An In Vivo Study on Hippocampal Oxidative Stress and Neurosteroids, Int. J. Mol. Sci. 23 (2022) 1365. https://doi.org/10.3390/ijms23031365

[36]   R. Wang, B. Song, J. Wu, Y. Zhang, A. Chen, L. Shao, Potential adverse effects of nanoparticles on the reproductive system, Int. J. Nanomedicine. 13 (2018) 8487–8506. https://doi.org/10.2147/IJN.S170723

[37]   O.A. Adebayo, O. Akinloye, O.A. Adaramoye, Cerium oxide nanoparticle elicits oxidative stress, endocrine imbalance and lowers sperm characteristics in testes of balb/c mice, Andrologia. 50 (2018) e12920. https://doi.org/10.1111/and.12920

[38]   S. Muzammil, J. Neves Cruz, R. Mumtaz, I. Rasul, S. Hayat, M.A. Khan, A.M. Khan, M.U. Ijaz, R.R. Lima, M. Zubair, Effects of Drying Temperature and Solvents on In Vitro Diabetic Wound Healing Potential of Moringa oleifera Leaf Extracts, Molecules. 28 (2023). https://doi.org/10.3390/molecules28020710

[39]   V. Simon, C. Avet, V. Grange-Messent, R. Wargnier, C. Denoyelle, A. Pierre, J. Dairou, J.M. Dupret, J. Cohen-Tannoudji, Carbon black nanoparticles inhibit aromatase expression and estradiol secretion in human granulosa cells through the ERK1/2 pathway, Endocrinology. 158 (2017) 3200–3211. https://doi.org/10.1210/en.2017-00374

[40]   Y. Li, X. Song, W. Wang, L. Wang, Q. Yi, S. Jiang, Z. Jia, X. Du, L. Qiu, L. Song, The hematopoiesis in gill and its role in the immune response of Pacific oyster Crassostrea gigas against secondary challenge with Vibrio splendidus, Dev. Comp. Immunol. 71 (2017) 59–69. https://doi.org/10.1016/j.dci.2017.01.024

[41]   M.H. Sarfraz, M. Zubair, B. Aslam, A. Ashraf, M.H. Siddique, S. Hayat, J.N. Cruz, S. Muzammil, M. Khurshid, M.F. Sarfraz, A. Hashem, T.M. Dawoud, G.D. Avila-Quezada, E.F. Abd_Allah, Comparative analysis of phyto-fabricated chitosan, copper oxide, and chitosan-based CuO nanoparticles: antibacterial potential against Acinetobacter baumannii isolates and anticancer activity against HepG2 cell lines, Front. Microbiol. 14 (2023). https://doi.org/10.3389/fmicb.2023.1188743

[42]   L. Xuan, Z. Ju, M. Skonieczna, P.K. Zhou, R. Huang, Nanoparticles-induced potential toxicity on human health: Applications, toxicity mechanisms, and evaluation models, MedComm. 4 (2023). https://doi.org/10.1002/mco2.327

[43]   M. Bundschuh, J. Filser, S. Lüderwald, M.S. McKee, G. Metreveli, G.E. Schaumann, R. Schulz, S. Wagner, Nanoparticles in the environment: where do we come from, where do we go to?, Environ. Sci. Eur. 30 (2018). https://doi.org/10.1186/s12302-018-0132-6

[44]   A. Malakar, S.R. Kanel, C. Ray, D.D. Snow, M.N. Nadagouda, Nanomaterials in the environment, human exposure pathway, and health effects: A review, Sci. Total Environ. 759 (2021) 143470. https://doi.org/10.1016/j.scitotenv.2020.143470

[45]   L. Farsi, S. Sabzalipour, M. Khodadadi, N.J. Haghighi Fard, F. Jamali-Sheini, The Ecotoxicity of Nanoparticles Co2O3 and Fe2O3 on Daphnia magna in Freshwater, J. Water Chem. Technol. 43 (2021) 509–516. https://doi.org/10.3103/s1063455x21060023

[46]   S. Singh, Adverse effects of nanoparticles on human health and the environment, Antivir. Antimicrob. Coatings Based Funct. Nanomater. Des. Appl. Devices. (2023) 305–330. https://doi.org/10.1016/B978-0-323-91783-4.00016-4

[47]   L. Ferraro, F. Dicé, A. Postigliola, P. Valerio, Le pluralità identitarie tra bioetica e biodiritto: un confronto interdisciplinare, Plur. Identitarie Tra Bioet. e Biodiritto. (2020) 13–24. https://doi.org/10.4000/books.mimesis.918

[48]   R. Sparrow, The Social Impacts of Nanotechnology: an Ethical and Political Analysis, Ethics Nanotechnology, Geoengin. Clean Energy. (2020) 175–185. https://doi.org/10.4324/9781003075028-12

[49]   M. Sajeer P, A. Keerthi, M. Varma, Solid-state nanopore conductance modulation using integrated microheaters, Nanotechnology. 36 (2025) 265301. https://doi.org/10.1088/1361-6528/ade318

[50]   S. Biroudian, M. Abbasi, M. Kiani, Theoretical and practical principles on nanoethics: A narrative review article, Iran. J. Public Health. 48 (2019) 1760–1767. https://doi.org/10.18502/ijph.v48i10.3481

[51] A.M. Soltani, H. Pouypouy, Standardization and Regulations of Nanotechnology and Recent Government Policies Across the World on Nanomaterials, Adv. Phytonanotechnology From Synth. to Appl. (2019) 419–446. https://doi.org/10.1016/B978-0-12-815322-2.00020-1

[52] A. Baran, Nanotechnology: Legal and ethical issues, Eng. Manag. Prod. Serv. 8 (2016) 47–54. https://doi.org/10.1515/emj-2016-0005

[53] "Echa Przeszłości," Iii (2002), Echa Przesz. (2024). https://doi.org/10.31648/ep.10823

[54] Terms of reference for an EU Bee Partnership, EFSA Support. Publ. 15 (2018). https://doi.org/10.2903/sp.efsa.2018.en-1423

[55] A. Ahlbom, J. Bridges, R. de Seze, L. Hillert, J. Juutilainen, M.O. Mattsson, G. Neubauer, J. Schüz, M. Simko, K. Bromen, Possible effects of electromagnetic fields (EMF) on human health--opinion of the scientific committee on emerging and newly identified health risks (SCENIHR)., Toxicology. 246 (2008) 248–250. https://doi.org/10.1016/j.tox.2008.02.004

[56] V. Amenta, K. Aschberger, M. Arena, H. Bouwmeester, F. Botelho Moniz, P. Brandhoff, S. Gottardo, H.J.P. Marvin, A. Mech, L. Quiros Pesudo, H. Rauscher, R. Schoonjans, M.V. Vettori, S. Weigel, R.J. Peters, Regulatory aspects of nanotechnology in the agri/feed/food sector in EU and non-EU countries, Regul. Toxicol. Pharmacol. 73 (2015) 463–476. https://doi.org/10.1016/j.yrtph.2015.06.016

[57] Figure 1: EPA Level III ecoregions in the state of Texas (shapefile downloaded from https://www.epa.gov/eco-research/ecoregion-download-files-state-region-6)., (n.d.). https://doi.org/10.7717/peerj.3612/fig-1

[58] M.A. Hamburg, FDA's approach to regulation of products of nanotechnology, Science (80-.). 336 (2012) 299–300. https://doi.org/10.1126/science.1205441

[59] A. Klinke, M. Dreyer, O. Renn, A. Stirling, P. Van Zwanenberg, Precautionary risk regulation in European governance, J. Risk Res. 9 (2006) 373–392. https://doi.org/10.1080/13669870600715800

[60] D. Freestone, The earth summit's agenda for change: A plain language version of agenda 21 and the other rio agreements, Glob. Environ. Chang. 4 (1994) 87. https://doi.org/10.1016/0959-3780(94)90026-4

Keyword Index

About the Editor

Jorddy Neves Cruz is a professor and researcher in the multidisciplinary areas of chemistry and molecular modeling. In 2021, 2022, 2023 and 2024 he was included in the list of Eminent Researchers in the area of Pharmaceutical Sciences at Federal University of Pará (AD Scientific Index). In 2023 and 2024 he was included among the Top 10,000 Scientists in Latin America (AD Scientific Index). He works as an Editor for the journals Frontiers in Chemistry, PLOS One, PeerJ, Molecules, Discover Toxicology, Current Medicinal Chemistry, Frontiers in Oral Health, Evidence-Based Complementary and Alternative Medicine, Combinatorial Chemistry High Throughput Screening, Journal Computational Biophysics and Chemistry and Journal Medicine, in addition to being a Reviewer for 61 International Scientific Journals.

www.ingramcontent.com/pod-product-compliance
Lightning Source LLC
Chambersburg PA
CBHW071223210326
41597CB00016B/1926